Design and Applications of Hydrophilic Polyurethanes

Medical, Agricultural and Other Applications

T0187949

Design and Applications of Hydrophilic Polyurethanes

Medical, Agricultural and Other Applications

T. Thomson

Director, Main Street Technologies, LLP

CRC Press
Taylor & Francis Group
Boca Raton London New York

CRC Press is an imprint of the
Taylor & Francis Group, an **informa** business

CRC Press
Taylor & Francis Group
6000 Broken Sound Parkway NW, Suite 300
Boca Raton, FL 33487-2742

First issued in paperback 2019

© 2000 by Taylor & Francis Group, LLC
CRC Press is an imprint of Taylor & Francis Group, an Informa business

No claim to original U.S. Government works

ISBN-13: 978-1-56676-895-5 (hbk)
ISBN-13: 978-0-367-39863-7 (pbk)

Library of Congress Cataloging-in-Publication Data

Main entry under title:
    Design and Applications of Hydrophilic Polyurethanes: Medical,
    Agricultural and Other Applications

Visit the CRC Press Web site at www.crcpress.com

Library of Congress Card Number 00-102577

Visit the Taylor & Francis Web site at
http://www.taylorandfrancis.com

and the CRC Press Web site at
http://www.crcpress.com

Completing this work required the love and patience of a partner.
For this I thank my wife, Marguerite, who makes it all worthwhile.

Polyurethanes in various forms have taken a permanent position in the inventory of functional materials. From automobile bumpers to foam insulation, this simple chemistry has won a dominant position due to its durability, physical strength, cost, versatility and availability. It can be made into elastomers, which are strong enough to withstand an automobile collision and yet with only minor changes in chemistry and processing can be used to make cushions for furniture.

Figure 1 shows the various forms polyurethanes can take. In 1992, 5 billion pounds were produced.

For the most part, the variations mentioned in Figure 1 were developed to serve a physical purpose. For instance, low-density, rigid foams are used as insulation. Low-density, closed-cell foams are used as flotation devices. High-density elastomers are made into automobile bumpers or similar applications where high physical strength and flexibility are critical.

Figure 2 shows typical applications, differentiated by two of their most important parameters: density and durability.

The overwhelming volume of polyurethane is used because of its physical properties. As mentioned, these materials were developed to fill some physical purpose. Their density and/or their hardness essentially describes their function. We will use a quality system analysis to describe or design a new product. In the case of conventional polyurethanes, that design will focus on physical properties.

The subject of this book, however, is an alternative class of polyurethanes. Foams and elastomers made from hydrophilic prepolymers are used because of their chemistry, specifically, their ability to absorb or otherwise manage water. In this sense, these materials are used to some degree for the chemistry of polyurethanes.

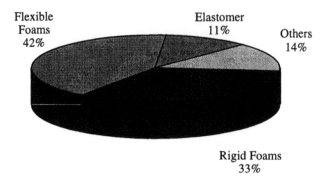

FIGURE 1. Distribution of the forms of polyurethanes (adapted from Reference [1]).

The most visible segment of this category is in hydrophilic poly-urethanes. Re-invented in the middle 1970s by W. R. Grace & Co., its function depends on its compatibility with water-based systems (it is hydrophilic as opposed to the bulk of the polyurethane world, which is hydrophobic). Applications range from cosmetics and personal care products to agricultural growing media to medical devices (absorbent dressings). The commonality in all these applications is that in each case there is a compelling reason for the material to be hydrophilic.

In the early 1980s, researchers began to notice that polyurethanes (both hydrophilic and hydrophobic) served well as a scaffold or entrapment of

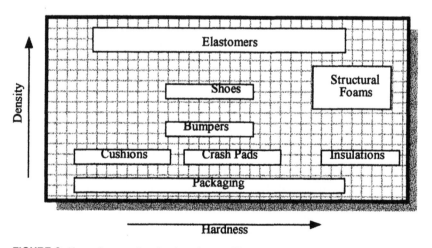

FIGURE 2. Uses of conventional polyurethanes differentiated by physical properties (adapted from Reference [2]).

biologicals. By these processes, living cells, yeasts, molds, algae and enzymes could be attached (immobilized) on the surface of, or imbedded in the matrix of polyurethane foams or films to create efficient means to produce enzymes or to degrade wastewater.

In this work we will describe the breadth of the work conducted in the area of hydrophilic polyurethanes over the past 20 years. Much of the work has been done with foams, but some references are made to elastomers. We begin with a review of polyurethane chemistry with an emphasis on the differences between conventional polyurethanes and hydrophilics. We will then present some notable examples of how this property has been used in commercial products. The final chapters of the book describe an infrastructure for the development of new devices. Topics include product development, processes, economics, quality systems, and analytical methods.

ACKNOWLEDGMENTS

My work in this area has been in three specific phases. The first was as the technical manager of the Hypol group with W. R. Grace & Co. My introduction to this technology was guided by Cliff Kehr and much of my education in the development of medical devices came as a result of collaborations with Hypol users.

The Hypol environment focused on research and development. It was my experience at Rynel Ltd., Inc., that taught me how to turn those bench-level developments into production volumes, with the requisite economic and quality considerations taken into account. During this period, I profited greatly from the counsel of Henry Jalbert.

Lastly, and in case I don't write another technical book, I want to express my thanks to Dr. L. Liefer, Professor of Chemistry, Michigan Technological University, for converting an immature, ignorant kid into something resembling a scientist. Apart from the knowledge that he forced into me, he taught me the discipline and life-style of the scientist. It was this foundation that gave me the skills to be able to take the information from Cliff and Henry and build it into the, hopefully, consistent story that follows.

CHAPTER 1

The Chemistry of Hydrophilic Polyurethanes

Polyurethanes are the result of the exothermal reaction between a polyisocyanate and a molecule containing two or more alcohol groups (-OH). Isocyanates contain a reactive group that can be made to react with -OH groups. Thus, if we react a diisocyanate with a molecule with an -OH group at each end, we create a basic building block of polyurethane; a prepolymer.

$$O=C=N-R-N=C=O \ + \ HO-(R'-0)_x-H$$

$$O=C=N-R-N-\underset{\underset{O}{\|}}{C}-(R'-O)-\underset{\underset{O}{\|}}{C}-N-R-N=C=O$$

SCHEME 1. Prepolymer reaction.

Notice that the ends of the prepolymer molecules are isocyanate groups. These are subsequently reacted to produce the solid foams and elastomers. As such, the material is usually a high-viscosity liquid. There are other ways to make polyurethanes, as we will discuss, but as we will show hydrophilic polyurethanes are made with the above prepolymer as an isolated intermediate.

All polyurethanes go through this reaction, but this step might be immediately followed by the reaction that produces the foam or other finished product. In the so-called one-shot process, the isocyanate and polyol, as well as catalysts, crosslinkers, surfactants and blowing agents are blended together in one step and deposited into a mold or other receptacle for the reaction. We will discuss this process later and compare it to the most common method for the production of hydrophilic polyurethanes, which is to produce the prepolymer and subsequently react it to make the final product.

Regardless of this processing variance, the chemistry of a polyurethane is sufficiently described by the contributions of each of the components. The isocyanate and the polyol are illustrated above but an additional component is necessary for the construction of a useful product, crosslinking. This component gives the product its durability.

Thus to understand the chemistry, we must deal with the effect of the following three components:

- the isocyanate
- the polyol
- the crosslinking

In as much as we will be describing a device that must provide an adsorbing surface (the biochemist's concern) as well as possess a durability to permit the construction of industrial equipment (a chemical engineer's concern), each of these factors must be understood and considered.

1.1 THE ISOCYANATE

The world of polyurethanes is predominantly split between two isocyanates, toluene diisocyanate and diphenylmethane diisocyanate.

Toluene Diisocyanate (TDI)

Diphenylmethane Diisocyanate (MDI)

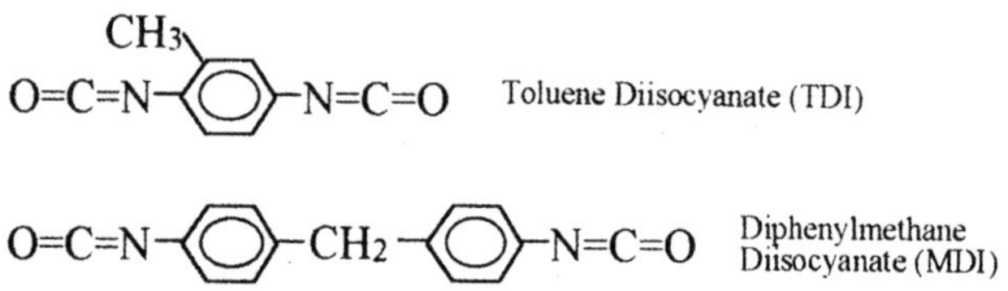

SCHEME 2. Aromatic diisocyanates.

Their relative importance depends on a number of factors. While TDI was the first successful isocyanate, and still remains important, it has been shown to cause cancer in laboratory animals, as have many isocyanates. Thus it is more difficult to handle from an industrial hygiene point of view.

Nevertheless, it is the preferred isocyanate for hydrophilics because it tends to make softer, more hydrophilic foam.

As seen from the following graph, the bulk of the polyurethane business has shifted toward MDI as the isocyanate of choice (see Figure 3).

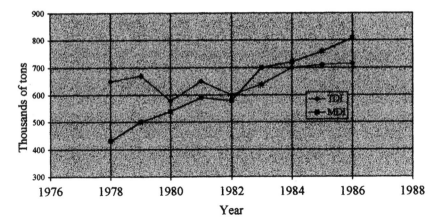

FIGURE 3. Growth of MDI versus TDI (adapted from Reference [3]).

Some hydrophilics are based on MDI but they tend to make more boardy foams. This is due, in part, to its increased mole percent in the urethane molecule. For a given polyol, an MDI-capped polyol has a higher percentage of hard segments and, therefore, is not as soft as a TDI-based polyurethane.

Another reason for the shift toward MDI is its faster reaction rate, which makes it more appropriate for the one-shot process mentioned earlier.

For a researcher involved in immobilization, the relative lengths of the isocyanate and the polyol are important. If the adsorption site were the polyol, one would want to minimize the weight percent isocyanate and focus on the proper polyol. As you will see there is a wide variety of polyols available.

While aromatic represent the dominant isocyanates in the conventional and hydrophilic polyurethane businesses, they present problems with respect to weathering, specifically yellowing on exposure to light and heat. While this may seem insignificant, the aesthetics of a product made from these materials are typically important. Whether the device is a cosmetic applicator or a wound dressing, yellowing is typically viewed as a degradation of the usefulness of the product. Thus, even if there is no evidence that the physical or hydrodynamic properties are affected by normal yellowing, it is almost always an issue.

Three processes cause the yellowing. Exposure to UV light causes the production of color bodies in aromatic isocyanates (TDI, MDI, etc.). This can be inhibited by the use of UV-absorbing compounds. Most commonly, however, packaging that is opaque to the ultraviolet is used to prevent yellowing.

Another major cause of yellowing is heat. Temperatures above 105°C can noticeably yellow foam in a few minutes.

Lastly, exposure to hydrocarbon emissions causes yellowing. For this reason, hydrophilic polyurethane foam manufacturers typically use electric forklift trucks. The effect seems to be related to the ability of the foams to strongly adsorb hydrocarbons on their surface.

When the yellowing has to be eliminated (as opposed to inhibited), other isocyanates are available. The most common are the aliphatics shown here.

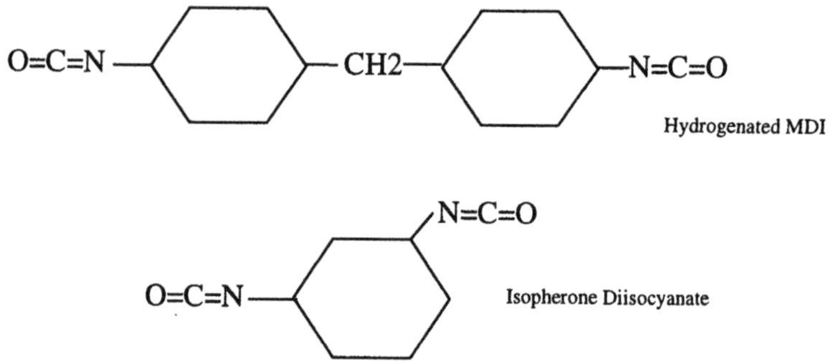

O=C=N —⬡— CH2—⬡—N=C=O

Hydrogenated MDI

O=C=N—⬡ N=C=O   Isopherone Diisocyanate

SCHEME 3. Aliphatic isocyanates.

While each of these has its own peculiar characteristics, both are aliphatic and therefore less subject to the yellowing that characterizes aromatic isocyanates. For instance, hydrogels made from isopherone diisocyanates are thought to have remarkably low protein adsorption characteristics.

1.2 THE POLYOL

For the most part, the polyol gives the polyurethane its chemical nature. Especially when TDI is the isocyanate, the polyol is the major constituent. Hence the secret to making soft foams is to change the length of the polyol chain.

Two types of polyols are typically used, polyesters and polyethers. The polyesters are usually based on adipic acid, but others are available. The polyethers are derivatives of ethylene and propylene oxides.

To follow are typical polyesters:

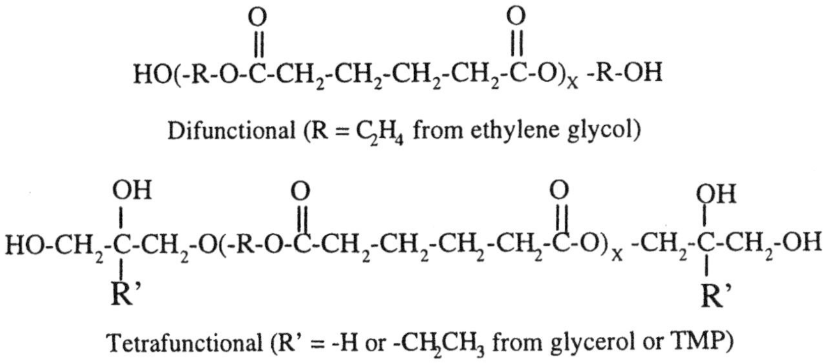

$$O \qquad\qquad O$$
$$\|\qquad\qquad\qquad \|$$
$$HO(-R-O-C-CH_2-CH_2-CH_2-CH_2-C-O)_X \ -R-OH$$

Difunctional (R = C_2H_4 from ethylene glycol)

$$OH \qquad\qquad O \qquad\qquad O \qquad\qquad OH$$
$$| \qquad\qquad\quad \| \qquad\qquad\qquad \| \qquad\qquad |$$
$$HO-CH_2-C-CH_2-O(-R-O-C-CH_2-CH_2-CH_2-CH_2-C-O)_X \ -CH_2-C-CH_2-OH$$
$$| \qquad\qquad\qquad\qquad\qquad\qquad\qquad\qquad\qquad\qquad |$$
$$R' \qquad\qquad\qquad\qquad\qquad\qquad\qquad\qquad\qquad\qquad R'$$

Tetrafunctional (R' = -H or -CH$_2$CH$_3$ from glycerol or TMP)

SCHEME 4. Polyols used in conventional polyurethanes.

Notice that these are essentially hydrophobic chemicals and therefore lead to hydrophobic polyurethanes. The structure of the polyethers is as follows:

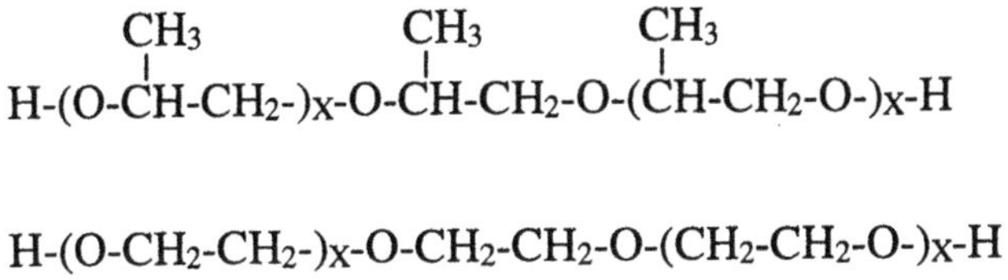

$$CH_3 \qquad\quad CH_3 \qquad\quad CH_3$$
$$| \qquad\qquad\quad | \qquad\qquad\quad |$$
$$H-(O-CH-CH_2-)_x-O-CH-CH_2-O-(CH-CH_2-O-)_x-H$$

$$H-(O-CH_2-CH_2-)_x-O-CH_2-CH_2-O-(CH_2-CH_2-O-)_x-H$$

SCHEME 5. Polyether polyols.

The propylene-based polyols are also hydrophobic and are the basis of most conventional polyurethanes. Polyethylene glycol is the basis polyol for most hydrophilic polyurethanes. Various molecular weights and the number of –OH groups make up the family of polyurethane-grade polyols.

In current practice, foam manufacturers prefer polyethers for the following reasons:

- lower cost
- better hydrolytic stability
- greater flexibility

1.3 CROSSLINKING

The last component of a polyurethane is the degree to which it is crosslinked. Without this component, the resultant polyurethane would not have the necessary physical strength to function. Making a stable foam, for instance, would be impossible.

One can identify specific isocyanates and polyols but specification of the degree of crosslinking is more problematic. Polyols can be used that include a crosslinking component. In this case the product developer chooses a polyol with more than two -OH groups. The physical effect of this choice on the foam is seen in Table 1. The average number of -OH groups can be chosen, and this can lead to a certain controlled amount of crosslinking. A component can be added to the prepolymer reaction to develop crosslinking, which has the effect of increasing the number of OH sites with which the isocyanate can react. This is typically the least expensive way to develop crosslinking.

The primary method of control, however, is the choice of the degree of functionality of the polyol (number of -OHs per molecule) whether this is done with a single polyol or by adding other, low-molecular-weight polyols. Table 1 shows the effect of functionality on the stiffness of the resultant foam.

The production of hydrophilic prepolymers involves selection of the amount and form of crosslinking. The addition of small amounts of water leads to the formation of urea linkages, which result in crosslinking. This method is typically not used due to the difficulty in controlling the reaction. More commonly, a low-molecular-weight polyol is added in controlled amounts. Trimethylol propane or glycerol is used. There is also a thermal crosslinking method that involves heating the reacting prepolymer to above 125°C. This results in ring opening and crosslinking.

1.4 SUMMARY OF THE COMPONENTS

The components above sufficiently describe the components of a polyurethane. Part of the universality of its application is rooted in the

TABLE 1. Physical Effect of Crosslinking [4].

Polyether Polyol Degree of Functionality	End Use
Linear PPG: functionality = 2.2+	Flexible foams/coatings
Branched PPG: functionality = 3 to 4	Hard foams and adhesives
Branched PPG: functionality = >4	Rigid foams

simplicity of its constituent parts. While limited, they offer the product designer enough flexibility to create products from shoe soles to chronic wound dressings.

It is now useful to describe how these components are combined to make foam products (cushions, etc.). As mentioned earlier, there are two dominant processes: the "One-Shot Process" and the "Prepolymer Process." Both are important relative to the subject of this book, so it is worthwhile to briefly describe them.

1.4.1 The One-Shot Process

In the late 1950s, it was discovered that one could manufacture a polyurethane directly from the component parts using certain surfactants and catalysts, thus avoiding the added step of a prepolymer process. In the one-shot technique, a polyol blend is made containing the surfactant, catalysts, blowing agents and other components. This blend is then quickly and intimately mixed with an isocyanate phase in what is called an impingement mixer. The emulsion that is created is then placed in a mold or other receptacle where the foaming reaction proceeds.

A typical formulation used in the one-shot process is shown in Table 2.

While an efficient process, due to the nature of the chemistries and formulations involved, no hydrophilic polyurethanes are made by this technique.

1.4.2 Prepolymer Process

In the prepolymer process, the isocyanate, the polyol and the crosslinking agent are reacted and isolated as intermediates. Prepolymer can be purchased of various constituents. Hydrophilic polyurethanes are all made from prepolymers.

TABLE 2. One-Shot Process Components (adapted from Reference [5]).

Component	Parts
Polypropylene glycol	100
Silicone surfactant	1.0
Water	4.5
Blowing agent	5.0
Stannous octoate	0.2
Dabco TL	0.1
TDI	115

The prepolymers are subsequently reacted with an additional component, or components, to yield the finished product. Most commonly the prepolymer is reacted with water to yield a foam according to the following chemistry.

$$O=C-N-R-N-C=O + HOH \longrightarrow O=C-N-R-NH_2 + CO_2 \uparrow$$

Foaming

$$O=C-N-R-NH_2 + O=C-N-R-N-C=O \longrightarrow$$

$$O=C-N-R-NH-\underset{\underset{O}{\|}}{C}-NH-R-N-C=O$$

Gelation

SCHEME 6. Reaction of a Hydrophilic Polyurethane and Water.

The first step is the reaction of water and the isocyanate. This yields the CO_2, which produces the foaming action. The other product of the reaction is an amine, which subsequently reacts with any available isocyanate to polymerize and eventually solidify the mass. The foaming and polymerization reactions must proceed at roughly the same rate so as to trap the CO_2 in the foam.

The prepolymer process is described in Figure 4.

Table 3 shows a typical hydrophilic polyurethane formulation.

While both processes are important to the researcher investigating the immobilization phenomenon, the prepolymer process provides slightly

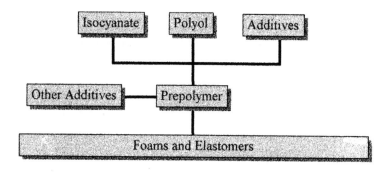

FIGURE 4. Schematic of the reactions to make a hydrophilic polyurethane foam.

TABLE 3. Hydrophilic Polyurethane
Foam Formulation.

Toluene diisocyanate	1 Part
Ethylene glycol (1000 MW)	2 Parts
Triol (optional)	0.01–0.05 Parts
Heat to 60°C to 120°C from 2 to 24 hours.	

more versatility. Because it offers the possibility of entrapping as well as adsorbing biologicals on the surface, one can envision greater possibilities.

This is especially true of hydrophilic polyurethanes. One typically mixes one part polyurethane with up to two parts of an aqueous solution (the aqueous). This aqueous can include a wide variety of constituents, including biologicals, nutrients and fillers.

While the mixing can be done by hand, a more uniform product is made from commercially available mixers.

1.5 CONCLUSION

Polyurethanes are in wide use because of their physical properties. Their chemistry, however, has been shown to offer the biochemist a friendly surface on which to grow cells or otherwise immobilize biologicals, including enzymes.

We have shown that the researcher has available a wide variety of chemical backbones including polyester, polyethers and isocyanates. Further, the degree to which the polyurethane is crosslinked offers additional variables.

Lastly, we have shown the processes by which these chemicals are combined. Special emphasis was put on hydrophilic polyurethanes, which offer a compatible surface for the aqueous-based systems that are common to biochemical research.

CHAPTER 2

Case Studies

In this chapter we will discuss a number of notable examples of how the special properties of hydrophilic polyurethanes have been used. The studies were chosen first because of their importance but, secondly, because the design features of the device were of sufficient interest to make them of technical interest.

In these discussions and in many of the chapters to follow, we will use a model for the development process. This model is described by the diagram in Figure 5.

The diagram seeks to emphasize the roles of each of the parties in the commercialization of a new product. The center portion represents the responsibilities of the product designer. That may be an individual or a committee of individuals in a company. The entity is responsible not only for defining what the product will do, but also sets bounds with respect to economics, toxicity, regulatory matters or any other factors that could affect the safety or efficacy of the device. In most instances, the product designer ultimately markets the device and in those cases, it is the designer's responsibility to include a differentiating factor to enhance the merchantability of the product.

Once defined, the product must be converted into a working device. This is represented in the left-hand portion of the diagram. The organization is responsible for the reduction of the design features into a working model. This may be the company chosen by the designer to manufacture the device or it may be an organization equipped to make prototypes only. In any case, the organization will use its expertise to fabricate a device that functions within the requirements of the design. Its responsibility does not stop there, however. As with any manufacturing process, cost and quality considerations must be taken into account. If the product is a medical

11

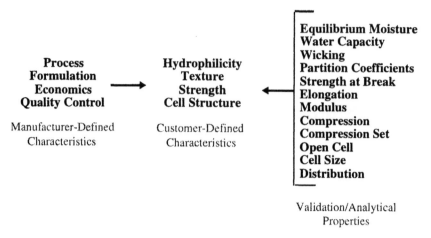

FIGURE 5. Product development model.

device, the manufacturer must fulfill certain record-keeping requirements. If the manufacturer is registered as an ISO 9000 facility, they will have to build a quality system concurrent with the development work.

This brings us to the last part of the diagram, the analytical or validation section. Especially for medical devices, but increasingly also for other products, it is necessary for both the designer and the manufacturer to assure the "authorities" that the product meets certain requirements. These include the design requirements, especially if they are included in a registration document, and that the product is safe and that the process by which it is made is under control. This is typically called a validation. Although this is not always possible to achieve, it is best that the tests used to evaluate the product are standardized. In those instances where specialized tests are required, the tests themselves need to be validated vis-à-vis their ability to represent an important feature of the device.

Thus the diagram seeks to describe the development process and the responsibilities of the parties involved.

In the following case studies, we will use this model to help the reader to see the commonality between the diverse devices we will discuss.

2.1 CASE STUDY: WOUND CARE

2.1.1 Introduction

The treatment of wounds in the United States is a multibillion dollar industry. In this sense wound is any breach of the dermis whether it is the result of a surgical procedure or an accident. The following list shows typical indications for a wound care product that might be included in a 510(k) (the FDA process to permit marketing a medical device).

- abrasions and lacerations
- dermal ulcers
- diabetic ulcers
- donor sites
- first and second degree burns
- venous stasis ulcers
- pressure sores
- superficial wounds
- surgical incisions
- vascular access sites
- partial/full thickness wounds

It is important to note that wound dressings don't heal wounds. The best we can expect from a dressing is that it creates an environment that is conducive to rapid healing. That is to say, the body has the capability to heal itself as long as certain nutritional and physical requirements are met. We will discuss these, but for many of the wounds above, little or no help is needed. In as much as hydrophilic polyurethane dressings are more expensive than gauze, for instance, it is only the most severe wounds that are subject to the benefits and special properties that a hydrophilic polyurethane dressing offers. By way of example, a surgical incision on a healthy patient will heal without the application of a dressing. Thus, even though a hydrophilic polyurethane dressing could be used, economics dictates that either no dressing or at most a gauze dressing is applied.

When there is a great amount of exudate (blood or other fluid), or when there is a cavity as a result of dead and missing tissue, more elaborate wound care is needed. It is here that the use of hydrophilic polyurethane dressings becomes indicated. Thus while the list above is comprehensive, when economic factors are included, the list reduces to:

- dermal ulcers
- diabetic ulcers
- donor sites
- first and second degree burns
- pressure sores
- partial/full thickness wounds

We will limit our discussion to the treatment of these wounds, using what has come to be the common term for these specific conditions, pressure sores.

The treatment protocol for these maladies involves two phases, prevention and treatment. Although prevention is not specifically the purpose of this book, it gives the reader a perspective that is useful.

2.1.2 Prevention

The flow chart in Figure 6 [6] represents the first phase in the identification of the risk factors contributing to the development of a pressure sore.

The first risk factor to be determined is mobility. The nature of the wound is such that if constant pressure is placed on a part of the body, especially the heal or coccyx or other bony protuberance, the blood supply is compromised and death and, ultimately, the loss of tissue result. Spinal cord injuries often result in the loss of mobility and are, therefore, a constant concern. If a patient is identified as having restricted mobility, education and physical assistance are advised.

The three major risk factors to be considered are: mobility, incontinence and nutrition. In terms of mobility deficiency, there are means to remove mechanical loading, including beds and chairs. Also, it is common practice to assist the patient in turning or otherwise avoiding constant pressure

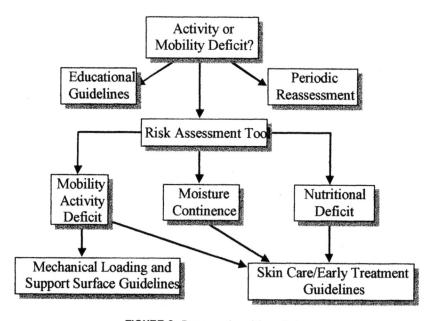

FIGURE 6. Pressure ulcer risk analysis.

on a specific part of the body. As we will discuss, once an ulcer is identified and dressed, removal of the pressure from the wound is an essential part of the treatment process. In as much as the pressure may have caused the problem, the treatment of the wound must include the dressing but removal of the cause.

As mentioned, the dressing that we apply to the wound is an aid to the body to heal the wound, i.e., the dressing does not do the healing. Accordingly, the caregiver that is assessing the risk factors must determine if the patient has the nutritional foundation necessary for the body to care for itself. Once an ulcer develops, the body must have the strength and raw materials to conduct the healing process. The healthcare community has made suggestions that need to be addressed both in the prevention mode and in healing.

Lastly, incontinence is a major contributing cause of the development of ulcer, especially in the coccyx area. Moisture causes a weakening of the skin (maceration). Once an ulcer forms, incontinence leads to a contamination of the wound site that can inhibit the healing process or cause a system infection. As we will discuss, special procedures and dressings are necessary if there is a strong likelihood of contamination of the wound by urine or feces.

2.1.3 Treatment

Despite precautions, ulcers do form, however. The wounds do not appear immediately as open wounds. The first stage is typically a reddening of the skin. The second stage is blistering. The third stage is evidenced by a loss of tissue. The depth ranges from the epidermis down into the dermis tissue (partial thickness wound). The fourth stage involves tissue loss down into the subcutaneous layers and even to the bone. Although each of these stages involves slightly different treatment protocols, the second third and fourth stages all involve some degree of tissue loss and need to be addressed specifically to encourage healing. The first stage is typically treated to prevent further damage using a simple film dressing.

The protocol in Figure 7 was developed as the first line of treatment of a skin ulcer [7].

Soon after the development of the ulcer, a crusty material composed of bodily fluids, blood and dead tissue develops. The first stage of the curing process is to remove this material. This process is called *debridement*. The wound is then cleaned using soaps and finally a saline solution. Once the root cause of the wound is identified, the wound needs to be dressed as part of the process to protect it from further injury and to encourage the healing process.

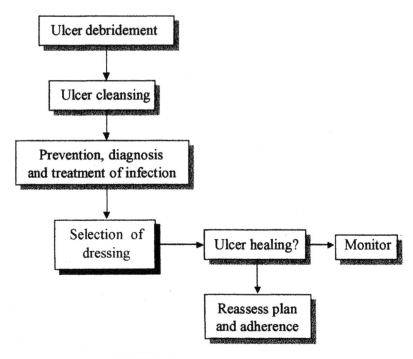

FIGURE 7. Treatment of a chronic wound.

There are a number of functions that a wound dressing must fulfill but the most important is to create an environment that encourages healing. Once the nutritional component and the mechanical factors are addressed, the dressing has a specific function. The function of the dressing in this respect was discussed by Winter [8]. The model in Figure 8 illustrates a newly debrided wound.

The epidermal and the dermal layers and the vasculature in the subcutaneous layer are shown. The wound is said to be a partial thickness because it extends into the dermis. If left unprotected, the wound will exude various fluids, which will dry up. The healing process begins with the development of new dermis tissue and the epidermis begins to develop. The epidermis, because of its nature, develops around the edge of the wound and grows inward. As can be seen in Figure 9, however, the drying out of the wound, while inhibiting the dermis, causes the epidermis to try to undermine the scab, thus inhibiting its growth. The scab is composed of fluids but also dried-up tissue. The scab itself is an attempt by the body to produce its own dressing. For simple wounds this is an effective process,

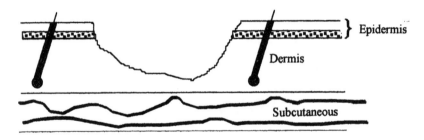

FIGURE 8. The partial thickness wound.

but for the wounds that are the subject of this case study, this process is not efficient.

This complication is avoided by using a wound dressing to serve as an artificial scab. The dressing, however, must have certain properties. As Winter noted, the primary property of the dressing is that it prevents or at least inhibits evaporation. By preventing the scab from forming, the growth of the epidermis is not retarded. Second, the new skin cells produced by the dermis will not dry out.

Winter describes others factors that are important to a good dressing. For example, he discusses what are called occlusive dressings, (Figure 10), which are meant to prevent air-borne bacterial contamination and other researchers have discussed gas exchange (CO_2 and O_2). The critical property, however, is to prevent the wound from drying out.

We are beginning to put together a list of requirements that a wound dressing would have if we are to design a new device. Factors include: protection, antisepsis, pressure, immobilization, auto-debridement, aesthetics, absorption, packing, support, information, cost, and comfort.

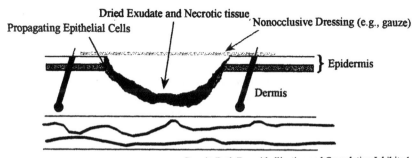

FIGURE 9. Treatment of a chronic wound with gauze.

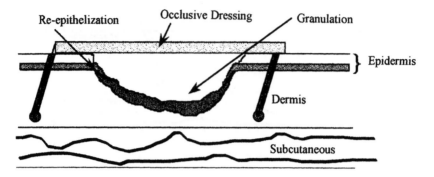

FIGURE 10. Treatment of the partial thickness wound with an occlusive dressing.

Speaking from the perspective of a product designer, I would include many if not all of these factors in my design, but I would also have to consider the market for the product. It is useful to take a few minutes to discuss this important factor.

Quoting from a multiclient study [9, p. 78], "Moist wound dressings came about as a realization that the use of gauze dressing which must be changed frequently and can damage new epithelial cells. Products introduced in the coming years will not only be less harmful to wounds, but they will also take more of an active role in expediting the healing process." The study projected that the moist wound dressing market was growing at a compound rate of 14.1% and would reach $467 mm by 1999. Those projections were essentially met. Continued growth through 2002 was expected.

2.1.4 Product Design

Referring to the above list, the design process begins with identification of the design features of a wound care product. This includes what problem is to be solved (refer to the above list of features) and may include how the problem is to be solved. Typically, the design addresses the commercial environment in which the new product will have to find its place. As an individual contemplates the introduction of a new product, within the context of the description of the technical and commercial market, a plan to cause the consumer buy will have to be included. This "approach to the market" is an additional design feature.

The next phase of the development process is the conversion of the design features into the formulation and process practices by which the dressing will be made. In the case of wound dressings, we will present one approach that could be used to fulfill the minimum requirements of a dressing. It is not an actual example but a formulation that could be used. Current manufacturers of foam wound dressings guard their formulations carefully and it would be inappropriate to use them as an example. We will, therefore, speak hypothetically.

The complexity of wound dressings based on hydrophilic polyurethane ranges from a simple foam pad to pads with an integral adhesive layer and what are called high-moisture vapor transmissive films. Since the commonality is the foam pad, we will discuss it further.

For purposes of toxicity and simplicity, most manufacturers and product designers have chosen the simplest of the hydrophilic polyurethane formulations to develop a dressing. The aqueous is composed of a simple, usually nonionic, emulsifier. The concentration is on the order of 0.1% of the aqueous, so the extractable are very low. Emulsifiers of various HLB ratings are used. As we will discuss in Chapter 3, the choice of emulsifier is determined by the properties of the foam desired. The following surfactants are known to be used in various commercial dressings: Pluronic F-88, BASF Corp., Pluronic F-68, BASF Corp., Pluronic L-62, BASF Corp., Brij 72, ICI Corp. and Brij 52, ICI Corp.

The application of an adhesive layer to a dressing has become an increasingly popular way to differentiate a dressing. When a foam pad is placed in the center of a fabric dressing such that the fabric overhangs the edges of the foam, the dressing is referred to as an island or composite dressing. In this case, the adhesive layer surrounds the foam pad.

When the adhesive is applied directly to the pad, care must be taken to ensure that the adhesive does not separate the foam from the wound and prevent the transfer of fluid into the foam. Several techniques have been developed to prevent this problem. In one procedure, a pattern or discontinuous layer of adhesive is applied to the foam. In another technique, the adhesive layer is applied in such a way as to produce a number of "pin holes."

Another way to differentiate, and some would say to improve the function of a dressing, is to cover it with a film that limits (but does not prevent) the evaporation of water. These films are known as high-MVTR (moisture vapor transmission rate) films. The rates at which they permit water vapor to pass through are expressed in units of grams/M^2/day. A number of manufacturers produce these films.

Application of these films to the foam, however, is not easy. Heat lamination or adhesive lamination techniques are used. Important considerations are the production of pin holes, which diminish the effectiveness of the film. Further, delamination from the foam when the foam gets wet is a constant development problem.

Since the foam swells when wet, application of the film presents problems. If the film is applied when the foam is wet, upon drying the film becomes wrinkled as the foam contracts. If the film is applied to dry foam, the resultant composite curls significantly when wetted.

To avoid this, several manufacturers apply the wet foam to a stretched film.

2.2 CASE STUDY: SPECIALIZED POSTSURGICAL DRESSINGS

We were asked to assist in the development of a specialized dressing. While it had all of what has come to be standard for a dressing, some attributes gave it some unique characteristics. Thus while the amount of absorption and the draining characteristics were high on the list of design requirements, three factors were included. These were:

(1) High tensile strength
(2) Low protein adsorption
(3) Nonuniform swelling

Thus a rather complex set of design requirements was developed, qualifying it for inclusion in this book.

2.2.1 Increasing the Strength of a Dressing

As we have said, the choice of hydrophilic polyurethane usually means that we sacrifice strength. In this case, however, the client was willing to decrease the absorptivity in exchange for physical strength. The dressing had to absorb, however, so hydrophilic polyurethane was the proper choice of material, but we were allowed some flexibility to diminish it. While in cases like this, increasing the crosslinking is a viable approach, the degree to which the strength had to be increased precluded a chemical approach to the development. Specifically, the finished dressing had to be able to withstand a tensile stress equivalent to hanging a 5 kilogram weight from it. Early tests showed that this could not be achieved without making the dressing hydrophobic, i.e., destroying its hydrophilicity.

Accordingly, the problem was approached from the perspective of adding a reinforcing fabric, typically called a scrim. The choice of materials, the size of the weave and, as it turned out, the type of weave were important.

With regard to the material, it is important that a strong bond be developed between the scrim and the hydrophilic polyurethane. It is not sufficient for this to be a physical bond. That is, the hydrophilic polyurethane must do more than flow into the fiber network of the fabric. The polyurethane must attach itself to the fibers and bond at that level. Otherwise the scrim will separate from the foam and the part will fail. Further, the fabric must be small enough not to interfere with the flexibility of the foam, yet still provide the physical strength to fulfill the design requirements.

The type of weave turned out to be the most interesting part of the project. Not being familiar with fabrics, we were pleasantly surprised to discover that one can find fabrics that expand in one direction and not in another. This turned out to be of great advantage. As mentioned, part of the design of the dressing was that it be of high tensile strength. The shape of the dressing was a rectangle, the length being about four times greater than the width. The tensile strength was necessary in the longitudinal direction, so a fabric that stretched minimally in that direction was required. As we have discussed, however, part of the absorption phenomenon is the swelling of the foam. To improve this it was desired that the foam be allowed to swell in the transversal direction. Besides improving the absorption, it also had the advantage that the swelling foam exerted enough pressure on the wound to produce the desired hemostatic effect (to stop bleeding).

The development effort thus focused on finding the right fabric. A polyester fabric was found that fulfilled these requirements.

Constructing the device presented its own problems. The goal was to embed the fabric into the foam. In order to prevent curling when the dressing became wet, it was important to get the fabric or scrim in the center of the device (Figure 11).

Several process variations were attempted. Placing the scrim in the center in the molding process was problematic and, as it turned out, unnecessary. Producing the product in continuous rolls was the most economical. Several options were investigated, but the easiest was to produce foam

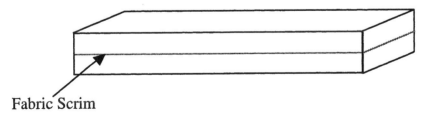

Fabric Scrim

FIGURE 11. Side view of the post-surgical dressing.

equal to half the thickness of the device. The product was then assembled as a sandwich, the fabric being placed between two foam layers. The layers were glued together using a prepolymer/aqueous emulsion.

2.2.2 Adsorption of Proteins

With regard to adsorption of proteins, it has long been known that hydrophilic polyurethane based on polyethylene glycol has low adsorption characteristics with regard to proteins. We will discuss the reason for this, with references, in the next chapter.

2.2.3 Nonuniform Swelling

Lastly, it was desired that any swelling of the dressing that took place upon absorption be limited to increases in thickness and not in length. This was accomplished and turned out to be an interesting feature of the dressing.

2.2.4 Other Design Features

While not part of the original design, several attributes were added to the project that gave the dressing particular marketability. These included taking advantage of the low compression set characteristic of hydrophilic polyurethane to produce a rolled-over edge. That gave the dressing a pseudo-three-dimensional (or molded) appearance even though the product was cut with standard steel-ruled blades.

Compression set is a characteristic of almost all foams and a standard test described in ASTM D-3574. It is a phenomenon described as a permanent deformation of the foam as a result of compression. Hydrophilic polyurethane foams are particularly subject to compression. When rolling foam after drying, therefore, care must be taken not to roll it too tightly as it will compressively set. The nature of the rolling process yields the unfortunate effect of compressing the center of the roll more than the outside. The result is a roll that, when unrolled, is thicker at the start than at the end. This problem is mitigated by aging the rolls before rolling and, as mentioned, rolling loosely.

Allowing the foam to age recognizes the fact the foam is typically not fully cured at the end of the process. Some experiments in our lab on the compression set phenomenon continues for several hours after the foam leaves the process. After several hours, it takes more pressure to achieve the same deformation. Going the other way in the process, before drying, it is easier to deform the foam. There is a point in the process called tack-

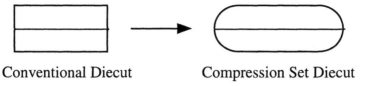

Conventional Diecut   Compression Set Diecut

FIGURE 12. Compression set phenomenon applied to diecutting.

free time (discussed in Chapter 4), which is the first point at which the foam can be handled. Applying practically any pressure at this point will permanently deform the foam.

In the case of this dressing, the compression set characteristic produced a dressing with rolled-over edges (Figure 12). This was considered both an aesthetic advantage and a clinical benefit.

2.2.5 Conclusion

The result was a dressing that fulfilled the original design features plus a number of other attributes that turned out to help differentiate the device.

2.3 CASE STUDY: AGRICULTURAL FOAMS

While outside the scope of this book, this development effort was unique in that it required control of the cell structure of the foam to a degree not typical for medical devices or other products. Of special interest is the way the cell structure of agricultural foam is measured.

The growth of a plant requires that the root system provide an opportunity for the plant to get both water and oxygen. If it is deprived of either, the plant will die. This is typically controlled by the void volume of the soil mix. In addition, the soil must be able to hold some quantity of moisture. While sand provides void volume, it does not hold moisture. Also, while clay holds moisture it does not permit exposure of the root system to oxygen.

For this reason, researchers have developed a number of commercial growing blends that include hydrophiles, like peat moss. Additionally, they include materials that seek to control the void volume, e.g., vermiculite, perlite, etc.

Early in the development of hydrophilic polyurethane foam, it was hypothesized that the material had the basic properties needed for a successful growing medium. Thus a business grew up around the use of this material. A weakness in the system was quickly realized, however. The competition for the foam was dirt. For this reason, while using pure foam could be justified, technically, it was necessary to use foam filled with large amounts of fillers. Most commonly, the filler of choice is peat moss, but bark and almost all natural fibers are also used.

Even with the cost-cutting method of adding fillers, however, hydrophilic polyurethane foam as a growing medium is only used with high-value plants. Among these are geraniums, cut flowers, roses, grapevines and a number of other plants. Billions of plants are grown worldwide using this technology.

However, the importance to us is not the commercial aspects but how the foam is built and the analysis of the final product.

There are a number of processes available and Figure 13 represents an average of the existing methods.

A slurry of peat moss (or other fiber) is made and an emulsifying agent is added. The choice of emulsifying agent is critical. First and foremost, it must not be phytotoxic. Second, it must yield an open-celled foam. As we have and will discuss further, the degree to which it is open is an important property. Unlike a wound dressing, the foam needs to drain. The explanation of this, and the method by which it is measured, is the most critical part of this technology.

Consider the following experiment. An apparatus is set up according to the diagram in Figure 14. Foam is placed in the reservoir (the volume of

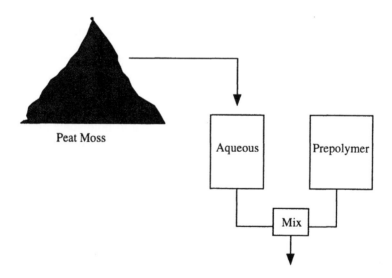

FIGURE 13. Process to make agricultural foam.

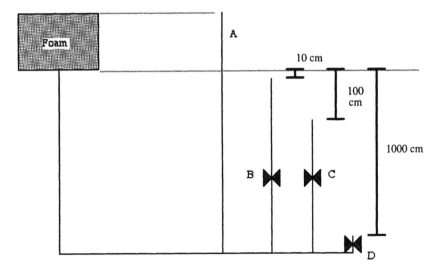

FIGURE 14. Equipment to measure moisture tension.

which is known) and is hydraulically filled with water as evidenced by liquid coming out of the tube A. This may take up to 24 hours. The weight of the foam is determined. Valve B is opened and water drains. When the draining stops, the foam is reweighed and the void volume is calculated as the weight of the lost water times the density of the water. Then valve C is opened, and the void volume is calculated again. This is followed by the opening of D and another calculation.

The data are reported as a graph of the void volume (*y*-axis) against the distance between the bottom of the reservoir and the top of the respective tubes (1 cm, 10 cm and 100 cm). Traditionally, the exponent (base 10) of the distance is used, i.e., 0, 1 and 2. The value is known as the moisture tension. Figure 15 illustrates both the data and a typical range of formulations.

You will notice that we are measuring the ability of the foam to drain. In the development of a wound care product where draining is considered a negative, the values from a similar analysis of Figure 15 would be off this scale. Thus all of the formulations would be considered highly draining in a wound dressing context. The formulations that are illustrated above show the ability not only to produce a draining form but to offer the possibility to finely tune the draining characteristics. It is felt that each plant requires a unique void volume/moisture ratio.

This has been the leading characteristic responsible for the growth of

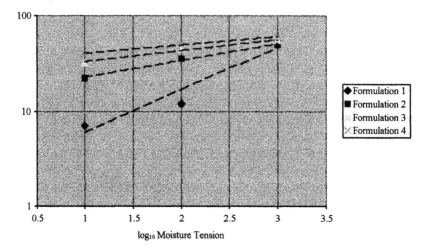

FIGURE 15. Moisture tension as a function of formulation.

this technology in agriculture. The differences in the formulations are the choice of emulsifier and the control of process conditions as we will discuss in Chapter 4.

2.4 CASE STUDIES: CONTROLLED RELEASE DEVICES

The technology of how one puts an "active" ingredient into a foam is covered in Chapter 4, but it is useful to go over a couple of examples here, if for no other reason than to discuss the calculations associated with the technology. In these examples we will not discuss rates of elution of therapeutic efficacy. Rather, once those values are established in clinical or other studies, how does one get an appropriate amount of active ingredient into the foam?

There are two basic methods to accomplish this and they will be addressed in detail when we discuss process. For now, we will simply describe them as in situ and imbibing. In the former technique, the active ingredient is included in the aqueous formulation. In the imbibing process, the foam is soaked in a solution of the active ingredient. While the in situ process is less complex and cheaper, it is not useable when the active ingredient reacts with the isocyanate. Typically, the active ingredient has some effect on the foam structure. The imbibing process has the advantage that in as much as the foam is created first, it can be specially designed for the application.

In this section, we will discuss two projects that required imbibing of an active ingredient into a foam. The processes to do this were sufficiently different to warrant their use as an example.

In both cases the active ingredient was soluble in the solvent used to imbibe the active ingredient. The solvent in these cases was water but it doesn't have to be. Practically, any solvent can be used since the foam is impervious to solvent attack. Only peroxides attack the foam.

Two methods will be discussed.

2.4.1 Method 1: Saturated Foam

A client asked us to scale up a process that is described as follows. A specially designed foam is saturated with an imbibing solution and is then carefully transferred to a tray and freeze-dried. The desired concentration of active ingredient in the foam was 200 μg/mg (20%).

The essential attribute of this method of imbibing is that the foam is saturated (at its fluid capacity, see Chapter 6). Thus, the calculation is based on the total volume of liquid in the sample as it is removed to the freeze-dryer. Refer to Table 4 for the method of calculation.

Upon drying, the sample will have 20% (or 20 μg/mg) concentration of the active ingredient. While this method works and is the basis of a very successful commercial product, it is inefficient because of the amount of water that must be removed. In defense of the method, however, it is said that this method of imbibing has specific advantages over the next method in that the active ingredient has a tendency to be closer to the surface of the foam matrix material.

However, this method has the potential disadvantage that the absolute value of the amount of liquid in the saturated foam is a function of the void volume of the foam and, therefore, a function of the density. Thus, in order to gain complete control of the process, one must know the density of the foam. This is alleviated by the next method.

Again, despite these notable problems, the design parameters of the device are fulfilled by the result of this process, a controlled delivery of the active ingredient. The other difficulties are, therefore, mute.

TABLE 4. Method of Calculation of an Imbibed Sample.

Wt. of foam sheet (grams)	25
Wt. of saturated foam (grams)	500
Wt. of liquid (grams)	475
Concentration of active ingredient in foam (%)	1.32%
% active ingredient in dried foam	20.0%

TABLE 5. Imbibing Calculation of EM Foam.

Wt. of dry foam (grams)	25
Wt. of saturated foam (grams)	500
Wt. of centrifuged foam (grams)	75
Wt. of moisture in foam (grams)	50
Concentration of active ingredient in water (%)	0.10%
Concentration of active ingredient in foam (%)	0.20%

2.4.2 Method 2: Equilibrium Moisture Method (EM)

In this instance, the client asked us to develop a process to add 0.2% of their active ingredient in a foam. We approached the problem from the perspective that using the equilibrium moisture was a more precise way of loading an active ingredient. To do so, we developed a process that rather than relying on the FC relied on what we felt was a more controllable factor, the equilibrium moisture. We will discuss this further, but for now we will concentrate on the solubility of the fluid (water) in the matrix material of the foam. It is therefore a function of the chemistry of the foam and not its structure (or density).

In this process the foam is saturated, and then the excess water is removed. In a process situation, this is most easily accomplished by centrifugation. By this technique virtually all but the water saturated into the matrix material is removed. The calculations are given in Table 5.

In practice it has been found that the loading a foam by the EM method is more precise. It is important to remember, however, that the method of manufacture is rarely a design requirement. The rate at which the active ingredient elutes is included in the list and is, therefore, the primary consideration.

2.5 CASE STUDY: ADSORPTION AND ENTRAPMENT OF BIOLOGICALS IN A HYDROPHILIC POLYURETHANE FOAM

In this section we will describe an emerging technology in the use of hydrophilic polyurethanes. While we will specifically describe this subject in terms of its use to bioremediate waste streams, the core technology is the immobilization (by adsorption, reaction or entrapment) of a biological species on or in a polyurethane foam. Included in these biological constituents are yeast, algae, bacteria, enzymes and even human cells.

The discussion begins with a description of the conventional treatment of municipal waste. It is common to use bacteria for this in large settling ponds. We will show how researchers have developed techniques to accomplish remediation of these waste streams by affixing the bacteria to a polyurethane substrate. The advantages are discussed.

We then discuss extensions of the technology to the treatment of other waste streams and to the production of chemicals.

The section concludes with a bibliography with abstracts of a number of research projects utilizing this technology.

2.5.1 Biological Treatment of Municipal Waste

Normally the treatment of liquid waste begins with the separation of solids. This is done typically in large separating ponds. Both continuous and batch filtration can also be used, as can centrifugation. The result is a dilute solution, the treatment of which is governed by the type of waste. Although there are a number of processes to accomplish the necessary purification, we are concerned with the process referred to as biological oxidation (biox).

The function of a biox system is to remove organic material from the wastewater by means of bacteria. The bacteria use the organics as food and metabolize them. The products of this metabolism are carbon dioxide, water, new cells and the bodies of dead cells. The goal is to remove oxygen-polluting components that, if left untreated, would find their way into the environment and have the effect of depleting the oxygen, which is detrimental to aquatic life. Although several physical and chemical processes exist to accomplish this, biological treatment is thought to be the most economical. Biox systems are essentially an accelerated form of the natural process to oxidize organic materials.

In order for the bacteria and higher microbiological species to proliferate and thus stabilize the organic waste, a suitable environment must be maintained. For aerobic bacteria, several conditions are critical, including:

- oxygen
- pH (ca. 7)
- availability of nitrogen and phosphorous
- absence of bactericides
- adequate mixing

Anaerobic bacteria, by definition, utilize materials other than oxygen, but for the purposes of this chapter we won't differentiate between the two. Certain rate-controlling variables that must also be considered include:

TABLE 6. Analysis of Wastewater.

Temperature (°C)	15
Dissolved oxygen (mg/liter)	2.0
BOD load (lb./1000 cu ft/day)	56
BOD load (lb./1000 cu ft/day)	113

Adapted from *Journal of Water Pollution Control Federation*, 1966, 38(6):939–956.

- microorganism concentration
- bacterial acclimation
- temperature
- contact time
- organic feed concentration

The processes by which biological oxidations are performed are similar. For illustrative purposes we will review the so-called aerated stabilization method. In this example, an influent was characterized by the analysis in Table 6.

The effluent was passed through a holding pond where it was held for about 4 hours. It was then pumped to the Aeration ponds that contained the bacterial soup. After about 66 hours of treatment, it was pumped to the settling pond where, after about 4 hours, the effluent was separated from the sludge, which was transferred to sludge storage (see Figure 16). The analysis of the effluent yields the data in Table 7.

Anaerobic ponds are used to treat low volumes of wastewater when insufficient land is available for aerobic oxidation. In some societies, anaer-

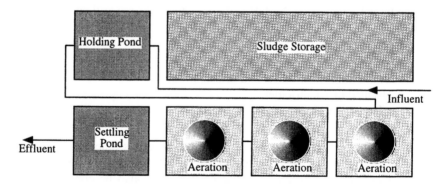

FIGURE 16. An aerated stabilization system.

TABLE 7. Analysis of Wastewater

BOD removed (lbs./day)	52,000
BOD removal efficiency (%)	48
COD removal (lbs./day)	86,000
COD removal efficiency (%)	39

obic systems are used to generate fuel for heating and cooking. For treatment of municipal waste, however, the generation of foul odors is a significant problem. Relatively deep treatment ponds employ aerobic oxidation at the surface while anaerobic decomposition takes place in lower strata. Completely anaerobic ponds are very deep, characterized by high loadings of organic materials. A two-step process accomplishes decomposition. First is the oxidation of the material to low-molecular-weight acids. This is followed by a breakdown into methane and other hydrocarbons. Only specific organisms are capable of this final degradation (called methanogenics).

With that discussion we will leave the conventional world of the biological treatment of waste to discuss research that focuses on using the same bacteria. Rather than producing a slurry of the bacteria and using artificial means to agitate the ponds or tanks, the bacteria is immobilized or fixed on a surface and the effluent is caused to flow over it. We will show that polyurethanes are an attractive scaffold or substrate on which to fix the bacteria.

As we said in the beginning of this chapter, the scope of this technology is much broader than waste treatment. By means of this technique, you can take a living organism and fix it to hydrophilic foam substrate. This applies not only to fauna but also to flora as we will show. This is true up to and including human cells.

The purposes of these composites of living cells and a synthetic surface to which they are fixed range from the production of chemicals, including enzymes, to the production of artificial organs. An example of the latter is a paper on the proliferation of viable liver cells in the matrix of a hydrophilic polyurethane foam.

In a related technology we will show that enzymes can also be immobilized on a hydrophilic polyurethane without decreasing their activity.

Before we get to that we should start with the chemistries and physics of the foams we will discuss. The chemistries described in the first chapter give the researcher sufficient information to investigate both the adsorption and the entrapment of biologicals on or in polyurethanes. There is little doubt that, as we will show, the adsorption of biologicals on the surface of the foam matrix is controlled in part by the chemistry of that surface. The three components of a polyurethane are therefore of interest to those who must optimize and commercialize these technologies. The issue of entrapment in a hydrophilic polyurethane must be investigated in the context where a wide variety of component parts are possible.

A factor that was not covered in the previous chapter is the physical structure of the foam. In this category are so-called open-celled foams and reticulated foams. In open-celled foams, some of the "windows" between cells have burst during the foaming process or by some postproduction technique so as to create a foam that has the ability to become hydraulically filled with fluid. Compare this with a closed-cell foam in which the cell windows are essentially closed and, therefore, have trapped the CO_2 or blowing agent.

A reticulated foam is one in which the intercellular material completely collapses into the foam matrix during the foaming process to produce a foam structure that looks more like I-beams and struts than foam. These materials are typically used as filters because of the manufacturer's ability to closely control the cell sizes and their strength. Reticulated foam can be made from hydrophilic polyurethane prepolymers but not to the same degree of precision as conventional PU. Many of the references that follow this chapter use reticulated foams. Unless otherwise noted, these are conventional hydrophobic polyurethanes.

In the remainder of this chapter we will review how researchers have used polyurethanes to immobilize biologicals for the purpose of producing a desirable product, degrade hazardous waste and improve the economics of enzyme processes. We will summarize the processes by which the biologicals are bound to the polyurethanes. Then we will describe how researchers have built these materials into function reactor schemes. We chose a representative paper for each of these categories. These papers were not selected by any other criteria than that they are typical of the given category.

2.5.2 Immobilization on Reticulated Foams

Pflugmacker, et al. [10] describe a method to produce what they call a biofilm on a reticulated foam by first washing the foam (PUR 90/16, Bayer, Leverkusen, Germany) with ethanol and placing it in a column fermentor. The column was then autoclaved and filled with a sterile culture medium. The column was inoculated with *Citrobacter freunii*. Flow was established and recycled for 24 hours at 36°C. Growth medium was continuously pumped into the reactor until a biofilm became visually evident on the support.

The purpose of this system was to produce 1,3-propanediol from glycerol. Using this technique, 57 moles of propanediol were produced from 100 moles of glycerol at a rate of 8.2 grams/liter.

2.5.3 Immobilization Using Prepolymers

Immobilization using prepolymer is described by Bailliez, et al. [11]. A wide variety of prepolymers were investigated including TDI- and MDI-based hydrophilics. In their studies a concentrated slurry of *Botryococcus braunii* was mixed directly with the prepolymers in small batches. The reactions were complete in 10 minutes. The resultant foam was cut into small pieces and washed with fresh culture medium. They were then transferred to a flask and growth medium was added.

After 24 hours the viability of the cells was measured by respiration and photosynthesis activity. By this rather severe treatment, most of the organisms were deactivated. A notable exception was the immobilization in a hydrophilic prepolymer based on MDI as the isocyanate.

2.5.4 Immobilization of Enzymes

By this technique, an enzyme is covalently bound to a prepolymer such that it can properly be described as a randomized co-polymer of polyurethane and the enzyme. In order for it to function as intended, however, the active portion of the enzyme must be unencumbered by the polymer backbone, either chemically or sterically.

Hu, et al. [12] immobilized β-D-Galactosidase using a TDI-based hydrophilic prepolymer. The foam was made by the reaction of the prepolymer and an enzyme stock solution. The foam was kept below 20°C. Enzyme leakage was monitored and the loading calculated by material balance.

Conclusions were that this was a viable method for the treatment of milk to make it lactose free.

2.5.5 Reactor Systems

A typical reactor scheme is presented in Pflugmacher [10]. By this technique, the temperature and pH are controlled. By their description, it is a fixed-bed loop reactor (Figure 17).

The fluid to be treated (D) is pumped through the column of foam A, which is immersed in water bath B by use of pump E into the receiver (C). The fluid in the column is continuously agitated by use of pump F and the temperature is controlled by use of the water bath B. The residence time in A is carefully controlled. The pH of the fluid is monitored and maintained by use of the pH probe and control system G.

We have here described the chemistry of polyurethanes as it relates to immobilization, some typical immobilization techniques and a useful reactor system.

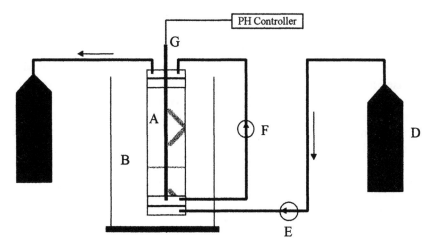

FIGURE 17. Bio-reactor.

2.5.6 Summary

This introduction is meant to give the reader a starting point from which to review the bibliography that follows. We have discussed the common practice of taking a municipal waste and treating with a bacterial soup to reduce its oxygen demand.

Lastly we have illustrated how some researchers have taken the properties of hydrophilic polyurethane to immobilize bacteria to accomplish the same task but more rapidly and efficiently.

The Appendix in this chapter is a compilation of research not only on this subject but on similar technology in practice. It encompasses the production of enzymes and hydrocarbons, treatment of other types of waste, and the development of medical devices, including artificial organs.

2.6 APPENDIX: EXAMPLES OF IMMOBILIZATION RESEARCH

Kinetic study of anaerobic digestion of fruit-processing wastewater in immobilized-cell bioreactors
Borja, R.; Banks, C.J.
Biotechnol. Appl. Biochem. 1994. Vol. 20, no. 1, pp. 79–92

Abstract: the kinetics of the anaerobic digestion of a fruit-processing wastewater [chemical oxygen demand (COD) = 5.1 g/l] were investigated. Laboratory experiments were carried out in bioreactors containing supports of different chemical composition and features, namely, bentonite and zeolite (aluminium silicates), sepiolite and saponite (magnesium silicates) and polyurethane foam, to which the microorganisms responsible for the process adhered. The influence of the support medium on the kinetics was compared with a control digester with suspended biomass. Assuming the overall anaerobic digestion process conforms to first-order kinetics, the specific rate constant, $K(0)$, was determined for each of the experimental reactors. The average values obtained were: 0.080 h-1 (bentonite); 0.103 h-1 (zeolite); 0.180 h-1 (sepiolite); 0.198 h-1 (saponite); 0.131 h-1 (polyurethane); and 0.037 h-1 (control). The results indicate that the support used to immobilize the microorganisms had a marked influence on the digestion process; the results were significant at the 95% confidence level. Methanogenic activity increased linearly with COD, with the saponite and sepiolite supports showing the highest values. The yield coefficient of methane was 270 ml of methane (under standard temperature and pressure conditions)/g of COD. The average elimination of COD was 89.5%.

Enhanced nutrient removal of immobilized activated sludge system
Shin, Hang Sik; Park, Hung Suck
Biotechnol. Lett. 1989. Vol. 11, no. 4, pp. 293–298

Abstract: an immobilized activated sludge system with porous polyurethane foam pads was operated in time-sequenced anoxic/oxic batch mode for enhanced nutrient removal. Biomass hold-up in polyurethane foam pads in the immobilized system increased with incoming organic substrate concentration. This new trouble-free system showed improved capability in nitrogen and phosphorus removal compared to conventional activated sludge system.

Immobilized cyanobacteria as a biofertilizer for rice crops
Kannaiyan, S.; Aruna, S.J.; Merina Prem Kumari, S.; Hall, D.O.
CO: 7. *Intl. Conference on Applied Algology,* Knysna (South Africa), Apr. 1996
J. Appl. Phycol. 1997. Vol. 9, no. 2, pp. 167–174

Abstract: N 2-fixing cyanobacteria (*Anabaena azollae,* symbiont strains) were immobilized in polyurethane foam and ammonia production by the cyanobacteria was investigated in the laboratory and rice fields. The cyanobacterial symbiont, *A. azollae*-MPK-SK-AM-24, showed the highest growth rate and biomass production among the five isolates examined while *A. azollae*-AS-DS showed the highest nitrogenase activity followed by *A. variabilis*-SA sub(0) (wild type, nonsymbiotic). Treatment of the foam-immobilized cyanobacteria with the systemic fungicide Bavistin stimulated nitrogenase activity while inhibiting glutamine synthetase (GS) activity. Free-living *A. azollae*-MPK-SK-AF-38, *A. azollae*-MPK-SK-AM-24 and *A. azollae*-MPK-SK-AM-27 excreted the highest amounts of ammonia into the growth medium; under foam-immobilized conditions the ammonia production increased further. Treatment of the foam-immobilized cyanobacteria with the fungicides Bavistin and Vitavax resulted in ammonia production at significantly higher rates. Rice seedlings (var. ADT 36) grown in the laboratory in conjunction with foam-immobilized *A. azollae* showed increased growth. A field experiment with paddy rice and foam-immobilized *A. azollae* strains indicated that the cyanobacteria excreted significant amounts of ammonia into the flood water in the rice fields, resulting in increased chlorophyll content of the plants and increased rice grain and straw yields. A combination of fertilizer nitrogen and inoculation with foam-immobilized cyanobacteria also significantly increased the rice grain and straw yield. Additionally, both *A. azollae* and *A. variabilis* were immobilized in sugarcane waste (bagasse) added to the rice paddy and resulted in increased rice grain yield.

Effect of the systemic fungicide Bavistin on the nitrogen status of cyanobacteria under immobilized state in polyurethane foam
Uma, D.; Kannaiyan, S.
S. Afr. J. Bot. 1996. Vol. 62, no. 3, pp. 127–132

Abstract: symbiotic forms of cyanobacterial cultures, namely, *Anabaena azollae* AS-DS and *A. azollae* SK sub(6), and the free-living forms, namely, *A. variabilis* SA sub(0), *A. variabilis* SA sub(1), *Nostoc muscorum* SK and *N. muscorum* DOH, were immobilized in polyurethane foam (PUF) and examined for ammonia excretion, growth and nitrogenase activity. Good growth was established in 3 weeks, after which the PU foam-immobilized cyanobacterial cultures were treated with Bavistin, a systemic fungicide, at 5 p.p.m. Ammonia excretion was enhanced by Bavistin, while glutamine synthetase activity was markedly inhibited. Bavistin treatment stimulated growth and nitrogenase activity of the cyanobacterial cultures. *A. azollae* AS-DS and *A. variabilis* SA sub(1) recorded a higher protein content with Bavistin treatment. The chlorophyll-a content of *A. variabilis* SA sub(0), *A. azollae* AS-DS, *N. muscorum* SK and *N. muscorum* DOH was significantly increased. *A. variabilis* SA sub(1) recorded higher C-allophycocyanin and C-phycoerythrin. Carotenoid content was increased in *A. variabilis* SA sub(1), *A. azollae* AS-DS, *A. azollae* SK sub(6) and *N. muscorum* SK. The cyanobacterial cultures *A. variabilis* SA sub(1) and *N. muscorum* SK have registered significantly higher total carbohydrate levels on treatment with Bavistin.

Manipulation, by nutrient limitation, of the biosynthetic activity of immobilized cells of Capsicum frutescens *Mill. cv. annuum*
Lindsey, K.
Planta. 1985. Vol. 165, no. 1, pp. 126–133

Abstract: the relationship between the synthesis and accumulation of protein and capsaicin was investigated in cultured cells of *Capsicum frutescens* Mill. cv. annuum immobilized in reticulate polyurethane. Cells were cultured in media containing reduced concentrations of essential nutrients in an attempt to manipulate the rates of protein synthesis. A relationship was observed between the intracellular nitrate concentration, the culture growth index and the incorporation of (super(14)C)phenylalanine into soluble protein—each of these factors was inversely related to the incorporation of a label into capsaicin and the total capsaicin content of the cultures.

Immobilization of Bacillus stearothermophilus *cells by entrapment in various matrices*
Manolov, R.J.; Kambourova, M.S.; Emanuilova, E.I.
Process Biochem. Vol. 30, no. 2, pp. 141–144

Abstract: Bacillus stearothermophilus G-82 cells were immobilized by entrapment in alginate, agarose, agar, carrageenan, polyurethane and polyacrylamide. Polyacrylamide (20%) was the best support for immobilization, providing relatively high pullulanase production with minimal appearance of free cells in the medium. Fifteen-fold elevation of specific enzyme activity in culture liquid was observed for the immobilized system.

Immobilization and culture of coffee (Coffea arabica L.) *cells on novel culture systems using a basket-shaped unit, "EGSTAR"*
Koge, K.; Orihara, Y.; Furuya, T.
Biotechnol. Tech. 1992. Vol. 6, no. 4, pp. 313–318

Abstract: a simple and new basket-shaped unit for agitation made of stainless steel (EGSTAR), in which immobilized coffee cells in CA-alginate gel beads were packed, was placed in a jar fermentor (System-1). This system allowed the plant cells to grow submersed in the unit even at high agitation speed (650 rpm). Only a small number of cells exited out of the EGSTAR. Most of the purine alkaloids produced were released into the medium. Suspended coffee cells in the jar fermentor were also possibly immobilized onto a polyurethane foam sheet fixed inside the net of the EGSTAR (System-2). The total cells in System-2 biotransformed theobromine to caffeine (77.9%). Other plant cell suspensions were also immobilized as efficiently as were the coffee cells. Thus, System-2 is a simple and convenient system for immobilization of plant cells to produce secondary metabolites.

Immobilization of Burkholderia cepacia *in polyurethane-based foams: embedding efficiency and effect on bacterial activity*
Domingo, J.W.S.; Radway, J.C.; Wilde, E.W.; Hermann, P.; Hazen, T.C.
J. Ind. Microbiol. Biotechnol. 1997. Vol. 18, no. 6, pp. 389–395

Abstract: immobilization of the trichloroethylene-degrading bacterium *Burkholderia cepacia* was evaluated using hydrophilic polyurethane foam. The influence of several foam-formulation parameters upon cell retention was examined. Surfactant type was a major determinant of retention; a lecithin-based compound retained more cells than pluronic- or silicone-based surfactants. Excessive amounts of surfactant led to increased washout of bacteria. Increasing the biomass concentration in the foam from 4.8 to 10.5% dry weight per wet weight of foam resulted in fewer cells being washed out. Embedding at reduced temperature did not significantly affect retention, while the use of a silane binding agent gave inconsistent results. The optimal formulation retained all but 0.2% of total embedded cells during passage of 2 l of water through columns containing 2 g of foam. All foam formulations tested reduced the culturability of embedded cells by several orders of magnitude, but O_2 consumption and CO_2 evolution rates of embedded cells were never less than 50% of those of free cells. Nutrient amendments stimulated an increase in cell volume and ribosomal activity in immobilized cells as indicated by hybridization studies using fluorescently labeled ribosomal probes. These results indicate that although immobilized cells were mostly nonculturable, they were metabolically active.

A comparison of two techniques (adsorption and entrapment) for immobilization of Aspergillus niger *in polyurethane foam*
Sanroman, A.; Pintado, J.; Lema, J.M.
Biotechnol. Tech. 1994. Vol. 8, no. 6, pp. 389–394

Abstract: Aspergillus niger was immobilized by adsorption and entrapment in polyurethane foams and the efficiency of retention capacity, citric acid productivity and the operational stability of a fluidized bed reactor were then compared. The adsorption technique was superior to the entrapment technique, and it was possible to obtain bioparticles capable of keeping their activity for more than 25 days.

Immobilization of the microalga Botryococcus braunii *in calcium alginate gels and polyurethane foams. Effect of immobilization on the alga metabolism-metabolite recovery by solvent extraction*

Largeau, C.; Bailliez, C.; Yang, L.W.; Frenz, J.; Casadevall, E.
CO: 4. *International Meeting of the Society for Applied Algology*, Villeneuve d'Ascq (France), 15–17 Sep. 1987
Algal Biotechnology. Stadler, T.; Karamanos, Y.; Mollion, J.; Morvan, H.; et al. eds. 1988. pp. 245–254

Abstract: in the frame of the production of metabolites from immobilized cultures, we tested a list of candidate solvents able to recover periodically intracellular hydrocarbons from the immobilized, hydrocarbon-rich microalga *Botryococcus braunii*, without affecting the primary and secondary metabolisms of the cells. On the grounds of low toxicity, high hydrocarbon recovery, easy product separation and possible solvent reuse, hexane was selected as being in this case the optimal solvent.

Colonization of polyurethane reticulated foam biomass support particle by methanogen species
Fynn, G.H.; Whitmore, T.N.
Biotechnol. Lett. 1982. Vol. 4, no. 9, pp. 577–582

Abstract: the ability of methanogen species to colonize reticulated polyurethane foam biomass support particles (BSP) in a continuous culture system using formate as the carbon source was investigated. Scanning electron micrograph evidence and biomass measurements indicate that two methanogen species effectively colonized within the matrix of the support particle. The freely suspended colonized BSP are resistant to washout, and comparison of methane output of the immobilized culture and liquid culture of the methanogens indicates the potential for process intensification of methane production.

Immobilization of yeast cells in polyurethane ionomers
Lorenz, O.; Haulena, F.; Rose, G.
Biotechnol. Bioeng. 1987. Vol. 29, no. 3, pp. 388–391

Abstract: one of the most promising methods for immobilization of biocatalysts is their entrapment into a polymeric hydrophilic matrix. Naturally occurring water-soluble polysaccharides and proteins are, in general, suitable for this purpose; however, for certain applications the mechanical and chemical stability of hydrogels of this type is insufficient. Therefore, increasing efforts have been made to develop synthetic polymers for this application. Polyurethanes (PURs) with hydrophilic properties have been studied by different authors. This communication reports on the immobilization of yeast cells in water-swollen amphiphilic anionic PURs of predominantly linear structure.

Immobilized enzyme particles prepared by radiation polymerization of polyurethane prepolymer
Kumakura, M.; Kaetsu, I.
Helv. Chim. Acta. 1983. Vol. 66, no. 8, pp. 2778–2783

Abstract: the immobilization of enzymes such as cellulase by radiation polymerization of dispersed polyurethane prepolymer was studied using tolylene-2,4-diisocyanate and 2-hydroxyethyl methacrylate. The polyurethane particles were obtained by the dispersion of polyurethane prepolymer followed by radiation polymerization, in which the enzyme was immobilized on its surface by covalent bonding. The particle diameter of immobilized enzyme particles varied with monomer concentration and composition. The enzymatic activity of immobilized enzyme particles varied with the temperature of dispersion and irradiation, and decreased with increasing particle diameter.

Preparation of porous polyurethane particles and their use in enzyme immobilization
Wang X; Ruckenstein E.
Biotechnol. Prog. 1993. Nov–Dec; 9(6):661–665
Jn. Biotechnology Progress, ISSN: 8756–7938

Abstract: porous polyurethane particles were prepared as follows: (1) two low-molecular-weight polymers, namely, poly[methylene(polyphenyl isocyanate)] and poly(propylene glycol), were mixed with stirring at room temperature and allowed to react. (2) The reacted mixture was dispersed with stirring in mineral oil containing small amounts of water, the catalyst dibutyltin dilaurate and $CaCO_3$ powder. In the presence of the catalyst, the reaction between the two polymers proceeded to completion. Small particles of polyurethane are thus formed, which contain mineral oil and $CaCO_3$ as porogens. The particles obtained, separated by filtration, were treated with a solution of HCl in order to generate additional pores, extracted with benzene to eliminate the mineral oil present in the pores, and finally subjected to drying and sieving. The particles were investigated by scanning electron microscopy (SEM), infrared (IR) spectroscopy and specific surface area measurements. Lipase from *Candida rugosa* was immobilized by adsorption on the porous polyurethane particles and crosslinked with glutaraldehyde to enhance the stability of the immobilization. The biocatalytic particles were used for hydrolysis of triacylglycerides. The high activity of the immobilized enzyme, which per enzyme molecule can be higher than that of the free enzyme, reveals that the porous polyurethane particles constitute excellent supports for lipase.

Polyurethane foam and a microbiological metabolizing system
Triolo, R.P.
CA: Scotfoam Corp., Eddystone, PA (USA). 1985
NT: US Cl. 435/41; Int. Cl. C12P 1/00, C12N 11/04, C08G 18/12
RN: US Patent 4,503,150

Abstract: a polyurethane foam was developed. A microbiological metabolic process assay was improved for use in utilizing the foam as a support structure for immobilized cells.

Immobilization support for biologicals
Parham, M.E.; Rudolph, J.L.
CA: W.R. Grace & Co. Conn., New York, NY (USA), 1988
NT: US Cl. 436/531; Int. Cl. G01N 33/545, 33/549
RN: US Patent 4,794,090

Abstract: an assay support matrix for an immobilized biologicals assay comprises a support matrix, which consists of a microporous membrane or particulate media having a protein nonadsorptive polyurethane polymeric coating. A bioaffinity agent was immobilized on said polymeric coating by passive adsorption.

The effect of immobilization on the course of biotransformation reactions by plant cells
Vanek, T.; Valterova, I.; Pospisilova, R.; Vaisar, T.
Biotechnol. Tech. 1994. Vol. 8, no. 5, pp. 289–294

Abstract: the influence of different immobilization methods on the biotransformation of verbenol by *Solanum aviculare* plants cells was studied. The biotransformation course was compared using free and immobilized cells. Immobilization techniques included entrapment in alginate, pectate and carrageenane gels, in polyurethane foam and on the surface of polyphenylenoxide.

Influence of immobilization procedure and salt environment on functional stability of chloroplast membranes: experimental data and numerical analysis
Thomasset, B.; Thomas, D.; Lortie, R.
Biotechnol. Bioeng. 1988. Vol. 32, no. 6, pp. 764–770

Abstract: the interactions between chloroplast membranes and their microenvironment within artificial matrices (albumin-glutaraldehyde matrix, polyurethane foam) were investigated. A statistical analysis achieved on the parameter values has allowed a quantitative assessment of the global behavior of immobilized chloroplast membranes. From the mathematical analysis of the experimental data, the authors demonstrate that citrate used in the reaction media prevents photoinactivation of the electron transfer chain whatever the nature of the matrix or the type of the reactor. The use of an albumin-glutaraldehyde matrix or an open reactor during the experiments also has allowed a better stabilization of the photosystems under operational conditions.

Mucilage acts to adhere cyanobacteria and cultured plant cells to biological and inert surfaces
Robins, R.J.; Hall, D.O.; Shi, D.-J.; Turner, R.J.; Rhodes, M.J.C.
FEMS Microbiol. Lett. 1986. Vol. 34, no. 2, pp. 155–160

Abstract: the surfaces of cells of several species of cyanobacteria were studied using low-temperature scanning electron microscopy (SEM), and have been shown to be covered in a layer of hydrated mucilage. This mucilage is observed in specimens of *Anabaena azollae* adhering to plant cells in their natural symbiotic niche (the cavity of the fronds of Azolla species) and in samples of the various species of cyanobacteria immobilized on polyurethane and polyvinyl support matrices. The mucilage appears to maintain the close contact observed between the cyanobacteria and these surfaces. Comparable films observed surrounding plant cells immobilized on similar polymeric surfaces are considered to be performing a similar function.

Effect of pore size and shape on the immobilization of coffee (Coffea arabica *L.*) *cells in porous matrices*
Koge, K.; Orihara, Y.; Furuya, T.
Appl. Microbiol. Biotechnol. 1992. Vol. 36, no. 4, pp. 452–455

Abstract: the pore size and shape of porous matrices were evaluated as to their effect on the immobilization efficiency in cultured coffee (*Coffea arabica* L.)/cells. A hydrophilic porous matrix (13–20 pores/25 mm) and reticulate polyurethane foam (30 pores/25 mm) indicated more efficient immobilization than the others, in small cubes [1 cm super(3) $\times$ 9] and a strip [1 $\times$ 1 $\times$ 9 cm super(3)] at the end of the fourth subculture. Among the large cubes [9 cm super(3)], the reticulate with the largest pore size (13 pores/25 mm) was the most advantageous for immobilization. In the strip-shaped matrices [1 $\times$ 1 $\times$ 9 cm super(3)], immobilization was the most efficient in spite of its lower surface area as compared to the small cubes, except for those with the largest pore size. The strip-shaped foams, which were fixed on the inside of the flask against shaking, were effective for immobilization.

CHAPTER 3

Product Development

In the last chapter we gave examples to show how the properties of hydrophilic polyurethanes can be used to produce medical devices. In each case, a detailed description of how the device was to function and, equally important, how it would be evaluated vis-à-vis its intended use was developed. An overriding theme of this work is that the design or definition phase of the project is to be carried out by medical device professionals. Those of us in hydrophilic polyurethane technology don't always qualify as medical device researchers. It is important that a partnership develop between the device designer and the materials scientist to ensure that the decisions made during the development stage take place under the direction of the medical device designer. The reason is obvious. It is presumed that there is a market for a device that fulfills a clinical purpose. Throughout the development, it was important to maintain an eye on the target of the project, regardless of the processing difficulty it entailed.

Figure 5, in Chapter 2, is meant to emphasize that the formulation and process development stages are subordinate to the design of the device.

In the next two chapters, we will describe how these design requirements are met by molding the properties of hydrophilic polyurethane. In this chapter we will discuss the formulations by which the device is made. In the next chapter we will show how these formulations are used to produce the product. In some cases, the formulation not only includes the selection of components, but is extended to include the thermodynamic conditions under which the product is made. As we will show, the quality of the emulsion is the primary determinant of the structure of the foam. The quality of the emulsion in turn is determined by the choice of emulsifying agent, the amount of mixing energy contributed by the mixer and

43

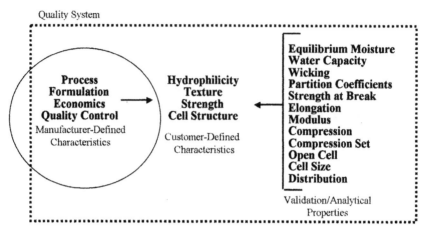

FIGURE 18. Process development stage.

the temperature of the constituents. All of these factors are considered, for the purposes of this work, to be part of the formulation.

Thus the product development structure (Figure 18) will be discussed in this and the next chapter as a sort of reduction to practice of the "invention" developed by the product design group.

During the presentation of the case studies, we discussed a number of physical properties that were part of the design properties that the project sought. Some were more important than others. Fluid absorption was considered to be the primary attribute of a wound care product, but as we showed, physical strength and conformability were also considered important. In many cases, it is these subordinate properties that serve as the basis for the marketing campaign. It is ultimately the marketing of the device that determines the commercial success. Very often the subordinate properties dominate how a product is differentiated from its competitors. In a recent very successful campaign, a wound care device was differentiated from other products by its nonuniform swelling characteristics. Other companies noticed this property and considered it to be an unfortunate characteristic of hydrophilic polyurethanes.

The following physical properties are considered to be the characteristics of hydrophilic polyurethane that are useful to the medical device designer. In this chapter we will give the reader an idea of how the material scientist begins the process of building the properties into a hydrophilic polyurethane-based product. Before we get to that, a word of caution: very few properties are independent. As we affect changes in cell size, we typically change the compression characteristics, the draining properties and a half a dozen other features. Thus, in most cases, the development process is the use of several techniques to affect a certain change. The goal is to solve the primary design problem while producing synergistic subordinate properties. Thus this chapter should not be considered as a recipe, but rather as a guide or starting place. Ideally, the reader will glean from the text the physical chemistries as opposed to using it as a formulary.

The properties we discussed in the case studies included (directly and indirectly):

- absorptivity
- strength
- adsorption of proteins
- cell growth
- permeability
- compression
- MVTR
- biodegradability
- elastomers
- density
- controlled release
- stiffness/conformability
- absorption under pressure

Each of the design features will be discussed with respect to how the material scientist will build that characteristic and how these properties are interrelated. As we go through this list, we will reiterate a theme that you will hear periodically. The primary feature of the device to be developed will, most often, include a requirement that the material be hydrophilic. This fulfills one of the screening concepts we discussed, that there be a compelling reason for the use of hydrophilic polyurethane. If, for instance, the primary characteristic were physical strength, in most cases hydrophilic polyurethane would not be the material of choice.

3.1 ABSORPTION

Absorption of fluids is the most prominent feature of hydrophilic polyurethane and is most often the primary reason for its use. Absorption is a more complex subject than one might think. How much fluid can be absorbed is a factor but, how strongly it is held, where it is held and how you get it out can have important ramifications for a product designer.

Each of these factors can be a positive attribute while others can be considered negative. In the next few paragraphs, we will discuss several aspects of absorption from how it is measured to how it is controlled. The goal is to give enough information to the product designer to use this important characteristic wisely.

The best way to describe the various facets of absorption is to describe how it is measured. Consider a piece of foam 4″ by 4″ by 0.25. The dry foam is weighed and placed in water at room temperature. The time it takes for the foam to visibly begin absorbing water is noted. Once that process begins, the water starts to fill the void volume of the foam. The time it takes to do that is of little importance, but for the next measurement to be meaningful, it is necessary for the foam to remain in the water until it is fully saturated. It is sometimes useful to squeeze the foam under water and allow it to expand under water. Ultimately, you will have to remove the foam taking care to reserve any water that drips from it and weigh it. Now return the foam to the water and allow it to absorb more water. This must be done until a constant weight is achieved.

This wet weight divided by the dry weight is called the fluid capacity (FC). For fully open-celled foam the density of the saturated foam approaches the density of water. We will discuss this more lately.

$$FC = \text{Saturated Weight/Dry Weight}$$

Now pick up the foam by a corner and allow water to drain until only the occasional drop falls off. Reweigh the foam. This weight divided by the dry weight is called the fluid capacity after draining (FC_d).

$$FC_d = \text{Drained Weight/Dry Weight}$$

Lastly, squeeze out as much water as possible by hand and then squeeze the foam again with paper towels. The goal is to physically remove as much water as possible. Once this is done, reweigh the foam. Using the following equation, calculate the last measure of hydrophilicity, the equilibrium moisture (EM):

$$EM = (\text{Wet Weight} - \text{Dry Weight})/\text{Wet Weight}$$

With that, our experiment is done, and we have described the major absorption characteristics of a foam. We will now examine each characteristic to describe its significance to the device designer. Following that we will show some of the methods by which these properties are controlled by the formulation and process.

The first phenomenon we observed was the rate at which the foam absorbs water. This is known by several names but the most common is *wicking*. A fast-wicking foam begins to absorb as soon as it is placed in water. Values of less than a second are usually reported as "<1 second." This can be an important technical factor and needs to be given attention during design. In an environment where significant amounts of fluid need to be controlled quickly, a fast-wicking foam becomes an important requirement. On occasion, fast wicking is an important marketing feature. I am thinking specifically of wound care. A highly exuding wound might put out 3–4 grams of fluid per day. This is slow enough to be easily handled by what might be considered a slow-wicking foam (e.g. 30 seconds). Nevertheless, the caregiver will most often pick the fast-wicking foam.

The control of wicking is strictly a function of the emulsifier package that is used. The foam itself has very slow wicking values (>5 minutes). If a water-soluble surfactant is used (Pluronic F-68, Tween 20, etc.), the wicking will be very fast. If one uses a waxy surfactant (Brij 72, for example), wicking will be relatively slow.

The cell size has some influence over the wicking but only at the high HLB values where interfacial tensions and contact angle inhibit the water from migrating into the cell structure.

Fluid capacity, the second value gathered in our experiment, is essentially the void volume minus the volume of the closed cells (usually very small for these foams). It is sometimes referred to as hydraulically filled foam. Consider a cubic foot of foam at a density of 6 lbs./cubic foot. The absolute weight of the polymer itself is 6 pounds. Its absolute specific gravity is about 0.97. The absolute volume is 0.1 cu. ft., leaving 0.9 cu. ft. of void volume. If this were a hydrophobic foam, the weight at its fluid capacity would be $(0.9 * 62.4) + 6 = 62.2$ lbs. and the FC would be calculated as $62.2/6 = 10.4$. But hydrophilic foam swells as it absorbs water into its matrix. As we will show, the amount it absorbs is measured as the equilibrium moisture, an important figure from a number of perspectives. For the purpose of this discussion, typical hydrophilic foam will double in value as it approaches EM. Thus for the cubic foot of dry foam we used in the above example, the wet volume is 2 cubic feet. The weight of the dry foam is the same but the swollen volume now holds an additional 62.4 pounds for a total of 124.6 pounds and a FC of 20.8.

Notice that if the density of the foam increases, the void volume decreases and the FC decreases. Also, as the EM decreases, the swelling and the FC decrease.

The FC, as such, is of little practical use. It is an indication of how open the cells are, but for most commercial applications, the value is of technical use only. The single exception was a process by which a pharmaceutical was imbibed into the foam. The foam was immersed in a solution of the drug. The water was then evaporated.

A more useful measure of absorption is the fluid capacity after draining (FC_d), although this should be used with caution. By this method, water is allowed to drain from the fully saturated foam by hanging it from a corner. This method is commonly used to specify a wound dressing. The philosophy is that a dressing is rarely used till it is hydraulically filled. When a dressing is removed, however, it is expected to hold onto the exudate strongly enough to minimize dripping. A leaking wound dressing, besides being unpleasant, would necessitate changing sheets, etc. Thus the method as we described it is an approximation of that effect. The caution arises from the fact that the test is meant to address a specific circumstance. That circumstance might not be appropriate for other applications. In those cases using one of the other measurements (FC or EM) might be advisable or a new, more appropriate test might even be devised. An example will follow that discusses the fluid capacity when the foam is under compression or is compressed (FC_c).

The control of this factor is with the process. As we discuss the process by which foam is made, we will describe a sequence of events that begins with the generation of CO_2 and gelation almost simultaneously. As long as the gel strength is able to withstand the internal pressure generated by the CO_2, the volume of the emulsion will increase. As the volume increases, however, the walls between the bubbles of expanding gas get increasingly thin. At some point, the strength of the polymer will be such that the pressure developed by the CO_2 will not be sufficient to distort the foam but begins to break the walls that separate the bubbles. This is the process by which open-celled foams are made. We will discuss this further in Chapter 6, on the foaming process, but for now we will stay with the effect on absorption.

A foam structure is made up of cells that are formed by the expanding CO_2. For the purposes of this discussion, let us say that each cell is surrounded by 10 other cells and that the separation is either a membrane (or window) or the residue from a burst membrane (i.e., open or closed). Consider the difference between a foam in which all the windows are broken and foam in which only 25% are broken. Both would be considered open-celled foams but the completely open structure would drain differently than the 25% foam. Put another way, the 25% foam would present more of a tortuous path for the water. Lastly, one could consider the cells in which many of the windows are closed to be small containers that hold water, preventing it from draining. This, coupled with surface tensions effect, has a dramatic effect on the FC_d.

From this discussion you can see the difficulty in applying this factor to a practical situation. Rarely would this phenomenon be of direct interest. Yet it is the most commonly used measure of absorption.

Control of the effect is a combination of formulation and process. A surfactant that produces a small cell (Pluronic L62 or Tween 20) offers a better chance at control. Emulsifiers that typically make larger cell ($>0.02''$) are difficult to control vis-à-vis FC_d. More important to the surfactant, however, are the measures used to control the relative rates of generation of CO_2 and the gelation. Temperature is the first line of control. If the temperature is too low ($<50°F$), the gel becomes too strong and the windows don't break. If the emulsion is too hot ($>90°F$), the cells become large and coalesce before a defined cell structure develops.

The difficulty is lessened, however, by proper application of an in-process test. In Chapter 6 of this book we discuss a technique known as air flow-through. This test measures either the pressure drop across a foam at constant flow rate or the flow rate at constant pressure drop. In either case it measures the difficulty air has in traversing the foam. In as much as this is directly affected by the number of closed windows, the measurement can be applied directly to the FC_d. The advantage is that the process can be controlled by this technique in real time.

Up till now the discussion of absorption could have addressed any foam regardless of hydrophilicity. Although the values of absorption would have been different, all foams, including so-called hydrophobics, can become hydraulically filled. They have control over the number of window that open and so some control of the draining is accomplished. All foams can be surface treated so wicking can be controlled. What separates hydrophilic foam from hydrophobics is the equilibrium moisture (EM).

EM is essentially the water that is absorbed into the matrix material. We discussed how it is measured but from a practical point of view its presence has two useful properties. First, it exists as a reservoir for active ingredients. Whether it is a pharmaceutical or soap, the active ingredient is buried deep in the foam matrix and is releasable when the foam is wetted. Second, as the matrix absorbs water it expands. Typically, a hydrophilic polyurethane doubles in volume when the matrix fully absorbs water. The water associated with FC does not have this property. By way of example, consider a 100 cc hydrophilic polyurethane foam that has an EM of 65%. The swelling will take place roughly according to the graph in Figure 19.

As the moisture increases, the volume of the foam increases, roughly linearly, until the EM is reached. After that there is no increase in the volume of the foam associated with an increase in moisture.

Control of EM goes back to the chemistry of the prepolymer. You will remember the three components of a polyurethane, the polyol, the iso-

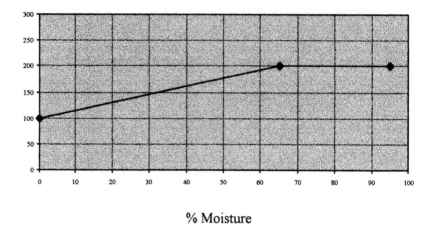

% Moisture

FIGURE 19. Swelling of a hydrophilic foam.

cyanate and the crosslinking. For all intents and purposes, all absorption is due to the polyol. As its concentration increases in the molecule, so does the EM.

Each of the commercial prepolymers has a unique EM that varies from 60 to 75%. Some manufacturers provide several grades of hydrophilic polyurethane prepolymers with different isocyanates and/or levels of crosslinking.

The product designer must be aware of EM from a number of perspectives. The first and foremost has to do with processing. When a formulation is developed, the amount of water in the aqueous phase should not exceed the amount that will be absorbed by the polymer matrix (the EM). If that guideline is not followed as the foam is processed, excess water has a tendency to separate from the foam. If this happens during the stage of the process while the prepolymer/aqueous emulsion is still a fluid (to be explained in Chapter 4), there will be a separation of phases that can and usually does destroy the quality of the emulsion.

Also of importance to the product designer is the swelling due to the EM. In many cases this is a benefit, adding to the overall absorptivity of the foam. But in some cases it is a disadvantage. In the case of wound care, for example, if a dry dressing absorbs exudate, it will swell and distort the pad. If this distortion is directed into the wound, one company with which we worked used that as a marketing advantage, commenting that the foam has a tendency to fill the wound. Other companies have sought to control the swelling by incorporating a fabric mesh, a polymer film, or both, to restrict the increase in the *x-y* directions. In pressure dressings, which are used primarily for leg ulcers where venous deficiencies are a problem, the increase in volume adds a certain amount of hemostatic pressure. This is desirable. This phenomenon is also used for the postsurgical dressing we discussed. That is, upon the absorption of fluids, the swelling exerts enough hemostatic pressure to control bleeding.

Lastly, the EM is the most precise way to control the addition of an active ingredient to the foam. Although each manufacturer and each product in each manufacturer's product line has a unique EM, variations with a single type of prepolymer are small. This is especially true when comparing EM with the other types of absorption. The other methods of absorption vary with cell size, degree of openness, etc. In as much as these are in part affected by variations in formulation and processing, EM is only a function of the chemistry and is, therefore, more uniform.

Another absorption phenomenon deserves attention. It is specifically intended to describe the efficacy of a wound dressing under compression. The treatment of leg ulcers, for instance, addresses the wound itself with dressings. The entire leg is commonly wrapped to apply pressure of up to 100 mm of Hg. The philosophy of this is beyond the scope of this book, but it is clear, as we have suggested in our examination of the way fluids are held in a foam, that compressing a foam dressing decreases its volume and therefore diminishes its ability to hold water. This diminished capacity, however, is limited to the fluid that is held in the cell structure of the foam, the fluid capacity. The equilibrium moisture is unaffected by the compression, as we will discuss.

A discussion of this property of hydrophilic polyurethane foam must include the foam's ability to resist compression. If we imagine a foam that is so stiff that it does not compress under a pressure of 100 mm Hg, the volume of the piece of foam would not change, nor would its FC. On the other hand, for a very soft foam, compression would decrease its volume significantly and therefore its FC would be seriously affected.

The apparatus pictured in Figure 20, was built to study this phenomenon. It is made up of a plate (1) attached to a screw mechanism (2), so that when the knob (3) is turned, the plate applies pressure to a foam sample (4). The foam is placed on an electronic balance (5) positioned under the pressure apparatus.

After zeroing the balance, the foam is compressed, and the force is read as an increase on the electronic balance. Knowing the area and the force, we can calculate the pressure (psi or Pa). By positioning a gauge (6) at some convenient place, the thickness of the foam sample can be determined. With this apparatus, the force that is required to compress a sample to a given thickness, or a percentage of the original thickness, is exerted. Conversely, the amount of compression experienced by a given pressure is determined.

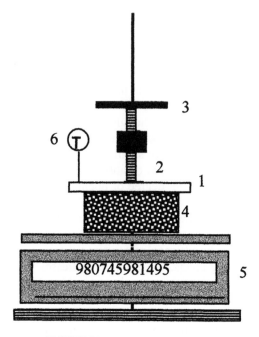

FIGURE 20. Compression apparatus.

Absorption of fluids under compression can be determined with the apparatus as well. If a fully saturated foam is placed in the apparatus (taking care to direct the fluid that will be expelled away from the electronics), the weight is recorded. The foam can then be compressed and the amount of fluid collected and weighed. By subtracting the weights, the amount of fluid that a foam under compression can hold is calculated. For fast-wicking foams, a dry foam can be placed in the apparatus and compressed. An eye dropper can then be used to saturate the foam with fluid. This is a more appropriate test for wound care although both techniques are useful. In the context of the technique, similar numbers are found for FG_c under compression.

Control of this property which, as we have discussed, is a function of the compressibility of the foam will be discussed below.

3.2 STRENGTH

As we have discussed, the choice of a hydrophilic polyurethane implies a decision that the strength of the device is not a primary consideration. The components of the polyurethane were not chosen with an eye to physical strength but to its relationship with the fluid in which it will operate. We just finished pointing out that one of the properties of a hydrophilic polyurethane is that it absorbs water into its matrix and swells to up to twice its volume. According to the equilibrium moisture concept, in this polymer matrix there is more water than polymer and, therefore, unless you believe in the "polywater" concept, lack of physical strength is an unavoidable characteristic.

All devices must have some strength; it just can't be the primary consideration. This puts strength as a property that is subject to optimization. The typical way to define this is that the device must be strong enough for the environment in which it must function. For instance, a wound dressing must be strong enough to be able to be removed without falling apart, even when fully hydrated. As an aside, the strength of the polymer decreases nearly linearly from zero to the equilibrium moisture. Additional moisture does not decrease the strength further. In other words, it follows the same curve as the swelling curve presented above.

While a discussion of strength is appropriate, it must occur in the context of the application and not relative to alternative chemistries, e.g., conventional polyurethanes.

The strength of the hydrophilic polyurethane is a function of two factors, one chemical and the other process. Chemical factors are familiar to all polymer chemists. As mentioned, the polymer is made up of three components. Since we are discussing hydrophilics, polyethylene glycol will be the backbone. Once hydrated, it has a strength that becomes the weak link in the polymer change. Thus when a hydrated chain is put under tension, the molecules will begin to break starting with the polyol. As these linkages break, the foam will stretch and continue to break until either the crosslinks or the isocyanate/urea linkages are put under tension. The point at which this takes place is determined by their concentration and in as much as these are stronger linkages (they are not hydrated to the same degree), it is clear that if we either increase the amount of isocyanate or increase the level of crosslinking, we increase the physical strength of the foam. Recognize, however, that if we increase the concentrations of either the isocyanate or the crosslinking, we also decrease the hydrophilicity.

In most cases, changing the chemistry is the last line of attack when additional strength is needed. The range of commercially available chemistries is not broad enough to provide any real flexibility in strength.

Another aspect of strength needs to be discussed. Referring to Figure 21, as the foam is put under tension, it will elongate to a point at which the molecular bonds begin to break. Typically, that point is reached at about 300% elongation. As stated, further elongation results in further breaking of bonds until the nonhydrophilic bonds are put under tension. Soon after that, the system fails and the foam physically breaks. This point

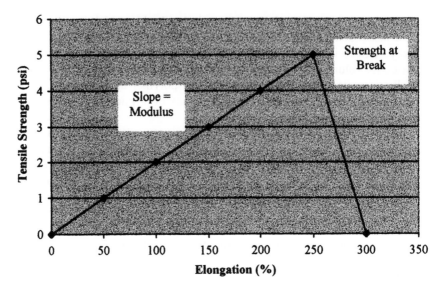

FIGURE 21. Strength and elongation of a foam.

is described as the "tensile strength at break," but an equally important number is the degree to which the foam stretched the elongation. This is, in many cases, as important a value as the tensile strength. Still further, the strength required to stretch an already elongated foam is called the modulus, and this is also an important factor. To illustrate, consider a wound dressing in place. To remove it, the corner is lifted and it is pulled to detach it from the wound site. The force required to detach it from the wound site is less than the tensile strength and typically less than the modulus. Thus, as the dressing is removed it stretches slightly and then separates from the wound surface.

In this situation, the tensile strength of the foam is not as important as the modulus.

As with tensile strength, both the chemistry and the process control elongation and modulus. Again the polyol backbone of the polymer is a fixed commodity for hydrophilic polyurethanes, so it is the relative concentrations of the isocyanate and the crosslinking that define strength. As the strength increases, the hydrophilicity goes down. For the same reasons, the elongation and the modulus go down (crosslinking and isocyanates are not elastic).

Although the chemistry factors given above are viable ways to control the strength, as a practical matter, there is an easier way. By controlling the density, one can affect small changes in strength without substantially changing the hydrophilicity or the elongation, although the modulus is increased.

Consider foam at 6 lbs./cu. ft. If the density is increased to 12 lbs./cu. ft., there is twice the mass of polymer in the same space. Thus the strength of the foam is proportionally higher. This is the most common way to increase the strength of a device. In most cases, this is the most effective way to develop physical strength. Remembering that the primary purpose is something other than strength, in most cases this is sufficient control.

There are occasions, however, when the strength of the foam is very important along with hydrophilicity. An example given in the case studies serves as a good model, but others could be mentioned. In the development of the postsurgical dressing, it was required that the dressing be able to withstand the pull of a 5-pound weight hung from a suture threaded through the dressing. Not only did the dressing have to withstand the tensile stress, it also had to resist the tearing produced by the suture. The foam was clearly not able to withstand that kind of treatment. Even changes in chemistry would not be sufficient. It was necessary to add a reinforcing fabric to the device. Commonly called a scrim, a process was developed that installed a fabric in the body of the foam. The scrim was strong enough to withstand the stress that was placed on it while not inhibiting the foam from absorbing fluids. The scrim prevented the foam from expanding to its natural swollen volume, so many of the absorption factors we discussed earlier (FC, etc.) were adversely affected, but the device was still able to absorb enough to function as required. In choosing a scrim, one must be careful to examine the bond between the foam and the fabric. The strength of the final device is to a great degree dependent on this bond. Cotton fabric (gauze) forms a covalent bond if it comes into contact with the foam emulsion. This is preferable. The use of adhesives to affect a good bond is also known.

In summary, strength should only be viewed in the context of what strength is required. If small changes in strength are needed, increasing the density is an option. If more strength is required, changes in chemistry can be considered but usually at the expense of hydrophilicity. If still more strength is needed, one might consider the use of a fabric scrim. In using a scrim, there must be good adhesion, preferably covalent bonding, between the foam and the scrim.

3.3 ADSORPTION OF PROTEIN

The adsorption of proteins on a hydrophilic polyurethane is a little studied effect from a fundamental point of view. We will cite a few researchers, but it is important to know that the phenomenon is at the molecular level. We have discussed the immobilization of biologicals (cells, yeasts, molds, etc.) on and in a hydrophilic polyurethane foam. Part of this is entrapment and part is a covalent bond formation. The present subject has to do with the interaction between proteinaceous material and a surface. Clearly, there is a relationship. It is also clear that the efficacy of these chemistries to serve as a scaffold or a template for cell growth is based on the adsorption phenomenon. In effect it is not too strong to distort cells, yet strong enough to provide a surface on which it can proliferate.

Braatz, et al. [13] studied the adsorption of proteins on polyurethane surfaces. They coated silica and silicone tubing with a variety of chemistries. Although they found differences, the polyurethanes demonstrated minimal protein adsorption given sufficient coating thickness. Correlations between the relative number of hard (isocyanate) and soft (the glycol) segments were drawn. More soft segments resulted in less adsorption.

Polyethylene glycols are well known as protein-compatible molecules when coated or grafted onto surfaces [14]. Both protein and platelet adsorption to PEG-modified surfaces has been shown to be reduced by PEG chains when attached to the surface at one end of the molecule. Adsorption and platelet attachment were shown to be inversely proportional to the length of the PEG chain, with 100 monomer units providing minimal adsorption and adherence [15].

A study of protein resistance of terminally attached PEG chains was performed [16,17]. In these theoretical studies, steric repulsion between the PEG chain and the approaching protein, as well as hydrophobic and van der Waals forces were calculated. High-surface density and long-chain PEGs were found to favor protein resistance.

Thus the control of adsorption is best addressed at the molecular level.

3.4 CELL GROWTH

In the case studies we reviewed the practice of biological oxidation of wastewater. We further showed how researchers have used the attributes of hydrophilic polyurethane to immobilize the same waste treatment sludges. Above, we discussed the interaction of proteins and the hydrophilic polyurethane surface and concluded that the PEG part of the molecule was the independent variable affecting adsorption. The question arises at this point, what is the connection between the two technologies? The answer is yet to be formulated. Taking into consideration the facts we do know, we have developed the hypothesis we mentioned earlier: the weak adsorption that proteins have toward the foam surface provides a template or scaffold on which cells can proliferate. The adsorption, however, is not so strong that the shape of the cells is not distorted. This is especially important when the cells must have a structure relative to one another in order to function, e.g., hepatic cells.

This reduces the control of this phenomenon to the relative concentration of the soft segments (the polyol) and the hard segments (the isocyanate).

3.5 PERMEABILITY/DIFFUSIVITY/CONTROLLED RELEASE

We discussed and illustrated the use of hydrophilic polyurethanes as controlled release devices. A number of factors control the rate at which an active ingredient exudes from a device. Among these are the solubility of the active ingredient, the adsorption of the solute to the hydrophilic polyurethane molecule, the molecular weight of the solute and the rate of diffusion (flux). Of these the diffusion is of particular importance with respect to control.

We want to explore the process of diffusion in some detail, but first I want to introduce a concept that simplifies how we look at hydrophilic polyurethanes. In the industry that has developed around controlled release, researchers differentiate between crosslinked water-soluble polymers (e.g., cellulosics, polyacrylamides, polyvinylperolidone, etc.) and hydrogels. This unnecessarily complicates the subject. In our research, we concentrate on the equilibrium moisture concept. If a polymer is insoluble in water and yet has an EM of greater than 10%, we define it as a hydrogel. This reduces the technology to a composite of a polar polymer with adsorption properties and water as a plasticizer. When using the system as a controlled release device, the solubility requirement is satisfied by the water. The higher the EM, the more solute the device will hold. Thus, we propose that hydrogels be rated with respect to their EM.

As discussed, the EM of a typical hydrophilic polyurethane foam is 60–70%. Chemistries have been developed that produce hydrophilic polyurethanes with an EM as high as 95%.

Let us now examine the diffusion process (Figure 22) with a goal of determining how one would affect changes in the release characteristics.

In order for this process to take place, two requirements must be fulfilled. A solute must be resident in the matrix material of the foam, and the partition coefficient must be such that the solute will leave the foam at therapeutic or effective levels. We will address these requirements individually.

There are two processes by which a solute can be placed into a foam.

Foam with Active Solute

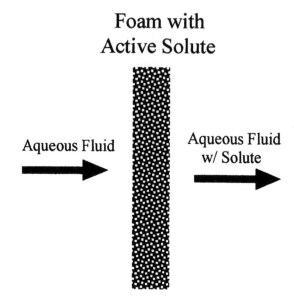

Aqueous Fluid

Aqueous Fluid w/ Solute

FIGURE 22. Diffusion through a foam.

These will be covered in some detail in Chapter 4. The first of these is to include the solute in the aqueous phase. This deeply embeds the active ingredient. The other technique is to soak the dry foam in a solution of the active ingredient. This is called imbibing.

In either case, the process depends on the active ingredient being water soluble. It is possible to put water-insoluble components into a foam. For example, an insoluble component can be dispersed in the aqueous phase. Alternatively, an alcohol- or alkane-soluble active ingredient can be imbibed into a foam. An example is imbibing a fragrance into a foam using ethanol. The result is a composite of an insoluble fragrance molecule in a matrix. If air is passed through this system, the fragrance volatilizes and a simple air freshener device is produced. As Figure 22 implies, however, we are dealing with components that are water soluble, so we will limit our discussion to components that will be released by an aqueous fluid passing through it. Thus the EM and the solubility of the active ingredient dictate how much solute will be resident in the foam.

We have discussed EM and how control is affected on a number of occasions. We know, for instance, that as the relative concentration of soft segments (the polyol) is increased, the EM increases. Thus, the amount of active ingredient that a hydrophilic polyurethane foam can hold is a function of the amount of polyol in the prepolymer.

We have also described how the active ingredient gets into the foam. To discuss how it comes out, we should examine the diffusion.

Once a solute is resident in a hydrogel, it can be released by placing it in an environment where there is a potential difference between the concentration in the gel and the concentration in the environment. By way of example, if a pharmaceutical is imbibed into a hydrophilic polyurethane foam, and the foam is placed into a device such that blood flows through it, the drug will leave the foam and enter the blood stream. The process will continue until a concentration is produced that is known as the partition coefficient.

$$P_c = [C_f]/[C_b]$$

where

C_f = concentration in the foam
C_b = concentration in the blood stream

The speed at which this takes place is a function of other factors, which we will cover shortly. The partition coefficient says that at equilibrium, the ratio of the concentration in the foam and the concentration in the blood stream is a constant. In as much as the blood is an independent variable in this equation (we can't affect significant changes in blood chemistry as a means of controlling drug delivery), the control of the partition coefficient rests with the chemistry of the foam. We have already discussed the adsorption of proteins on hydrophilic polyurethanes. Thus, in using a protein-based drug, the partition coefficient would tend to be higher than a drug with little or no adsorption effect.

The first line of control again leads us to consider the factors that affect adsorption, i.e., the polyol. The reader should, by now, realize that a number of diverse factors are controlled by the type and relative amount of polyol. This makes changes in chemistry a powerful tool but subject to a significant amount of optimization.

In most cases, it is the rate at which the solute is released that is of primary importance. Using the flow-through model pictured above, the rate at which the solute will be released to the aqueous stream is a function of the concentration difference in the foam and the fluid. The surface area and the diffusivity are also important. The last property is the rate at which the solute moves within the structure. Among other things, the rate at which the solute moves is a function of its molecular weight. For the purposes of this discussion, however, we will concentrate on the factors over which we have control. In this case, they are the relative differences in concentration and the surface area.

In the case of the release of a drug, the therapeutic and toxic concentrations need to be determined. With those and the partition coefficient, a concentration in the foam can be determined. The shape of the release curve is beyond the scope of this work; the reader should consult a text on controlled release vis-à-vis "zero-order release" patterns.

The structure of the foam is of prime importance, however, and its control will close out this section. An open-celled foam is the first requirement and many of the polyethylene oxide/polypropylene oxide block copolymers surfactants (Pluronics, Tween) surfactants serve well in this task. Apart from that, the control of surface area is affected through the amount of surfactant, the temperature and a number of other process-oriented parameters.

In summary, control of the release of a component to the environment begins with the chemistry of the prepolymer and extends to the formulation and process factors. These affect the most important dependent factors, the partition coefficient, the concentration and, finally, the cell structure and surface area.

3.6 COMPRESSION

In the analytical section of this book we will describe the quantitative compression characteristics of a foam, but for now we will think of compression as the sensory experience of touching or picking up a piece of foam. Although not typically a primary technical design consideration, there are applications where it is involved in the purchasing decision. As we know, the function of a wound care device is to maintain a moist environment and control exudate. Many of the wound care products on the market fulfill these requirements, and they are all identical in efficacy in those respects. At best they are indistinguishable from a practical point of view. Yet if a dressing whose compression characteristics are designed with the sensor impact as part of the design of the dressing, the consumer (nurses, in this case) will gravitate toward the dressing that "feels better." A softer dressing (a compression phenomenon) would appear to be more comfortable and conformable.

This is not to minimize the occasion when the compression characteristics of the foam are of actual technical interest. We discussed the use of hydrophilic polyurethane in a postsurgical dressing. In that case, and in similar devices, the expansion of the foam from dry to wet exerts a pressure on the tissue that can control bleeding. In this case, the force to compress the foam and the force exerted upon expansion can be considered the same.

The process to affect changes in the compression characteristics starts with the prepolymer. As with the other dependent variables we discussed, the polyol is not subject to change unless a lower equilibrium moisture is included within the design parameters of the device. If it is within the scope of the device to be able to lower the EM, then one might consider the use of more rigid polyols. A blend of PEG and a polyester might be advisable, for example.

Similarly, increasing the hard segments has the effect of increasing the force required to compress a foam. Increasing the TDI or switching to a blend of TDI and MDI (being careful to account for rate of reaction issues) is effective. Using MDI as the exclusive isocyanate yields a hydrophilic foam (if PEG 1000 is used) a very boardy foam with high compressive strength is produced.

The other hard segment is also an effective way to increase compressive strength, but again at the expense of hydrophilicity.

For a given prepolymer, formulation and the process can affect the compression characteristics. An aqueous that produces a larger emulsion size would have a tendency to make foam with larger cells and a corresponding intercell structure. This gives more structural integrity and more force to compress. Pluronic F-68 and Pluronic F-88 exhibit this characteristic. Nevertheless, the formulation can only be counted on to produce small changes in compressive strength.

A more dramatic effect is seen with density changes. An increase in density means that there is more polymer in the same space. It follows, therefore, that it will take more force to compress.

3.7 MVTR

The moisture vapor transmission rate (MVTR) is a measure of the rate at which moisture evaporates through a material. It is specifically intended to describe the ability of a wound dressing to inhibit evaporation from a wound site.

The process by which a dressing controls the evaporation from the wound is a complex, multifunctional process, much of which has nothing to do with the chemistry of the hydrophilic polyurethane foam. Nevertheless, once these other factors are normalized, the ability to control evaporation at the chemistry level presents itself.

In practice, however, this is not typically done. As it turns out, it is more appropriate to make composites of foam dressings with films of hydrophilic materials.

If we consider a cavity wound open to the air, the rate of evaporation is controlled by the humidity, temperature and air movement around the wound. If we cover the wound, some of those factors are eliminated (particularly, the movement of air), so the evaporation is diminished. If we cover the wound with gauze, as is common practice, we know the evaporation rate (given sufficient exudate) will be on the order of 10,000 grams/m^2/day.

If we had covered the wound with a Saran polymer film (which is not permeable to water vapor), the resultant evaporation rate would be near zero. Although this would fulfill the requirement to maintain a moist environment, it would result in "pooling" of the exudate, which would mascerate the wound causing further damage.

Rather than Saran, if we were to place a foam wound dressing, we would find that the rate of evaporation would increase to about 3000 grams/M^2/day.

The reason for the differences in rates (gauze to Saran to hydrophilic polyurethane foam) goes back to the permeability of material to water. The Saran cannot absorb water and therefore cannot transmit it, while gauze wicks water from the wound to the outside of the wound very quickly without absorbing it. The hydrophilic polyurethane, on the other hand, has a strong affinity for water and therefore builds up a reservoir as equilibrium moisture, which is tightly bound on a molecular level. A partition function is then set up between the hydrated hydrophilic polyurethane and the surrounding air. Once saturated, it is the diffusion of water in and out to the hydrophilic polyurethane that controls the rate of evaporation.

We have shown that we can affect this diffusion by changing the ratio of hard and soft segments (hydrophilic and nonhydrophilic segments) in the prepolymer. Refer to Section 3.5 of this chapter for the details.

It is interesting to note that the cell structure and size have little to do with the evaporation. At some point, there may be an effect but only when the moisture level has exceeded the EM level in the dressing. Up to that point, the water must diffuse through the matrix material. When the water exceeds the EM, larger cells are more subject to outside air currents and could conceivably increase evaporation over smaller cells. In practice, a wound dressing rarely (or at least should rarely) exceeds the EM.

We mentioned the typical value of 3000 grams/M^2/day value. In most cases this exceeds the rate at which a wound exudes. For instance, a typical wound of 25 cm^2 might produce 2 grams/day (800 grams/M^2/day). If the dressing evaporates faster than that, there is danger of drying out the wound. Control of evaporation above that provided by the absorbant foam is therefore called for. Typically, this is accomplished by laminating a film on the surface of the dressing away from the wound. There are reasons other than evaporation for applying a film. Putting an occlusive film on the dressing isolates the wound from bacterial communication with the atmosphere. It is also said to provide an exchange of CO_2 and oxygen across the film barrier. Lastly, and most importantly in my opinion, it provides for some degree of protection against contamination of the wound site from contact with urine and feces. Given the site at which many of the wounds occur and the prevalence of incontinence in these patients, this is not only a possible source of contamination it is often very likely.

There are a host of manufacturers of films that function appropriately in this application. For example, Bertek, Inc. of St. Albans, Vermont, has a number of films that exhibit various levels of permeability. In the graph in Figure 23 we illustrate some of the products.

Affixing the film to the foam is a task not to be taken lightly. An adhesive bond typically fails when the dressing is wetted. Heat laminating can be done but care must be taken not to create pin holes when the polymer sags into a cell. This destroys the waterproofing. Flame lamination, for some reason, does not seem to work with hydrophilic polyurethanes. The best results involve the use of both an adhesive and heat. Apparently, the adhesive bond and the physical bond are necessary. Be aware, however, that including an adhesive in the construction (even an acrylic adhesive) significantly lowers the MVTR.

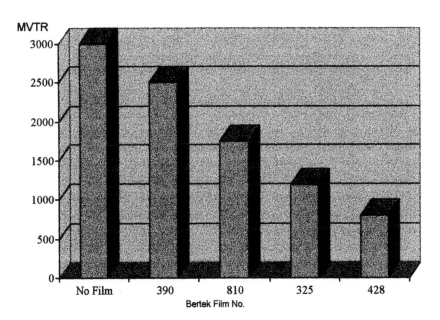

FIGURE 23. Evaporation rate through film/foam.

Alternatively, film and foam can be bound together during the foaming process. In this technique, a film coated with an adhesive is coated with the prepolymer emulsion and processed normally. In this case a covalent bond is produced that is very durable.

If this technique is used, the film has to be stretched before application of the foam. Otherwise, upon drying, the foam will shrink and cause severe wrinkling in the film, which is aesthetically unacceptable.

3.8 RESILIENCE

The rate at which a foam recovers its shape or height after being compressed is called its resilience. The measurement of resilience is covered in Chapter 6, but for now we are limiting ourselves to a discussion of its control.

The most important application for the control of resilience, and indeed one of the most important applications of hydrophilic polyurethane, is a sound-attenuating earplug.

Small changes in resilience can be designed into the foam at the prepolymer level but the magnitude is of little practical interest. We will discuss compression set below, and, as it turns out, the changes in chemistry that control that property are functional here also. It is a matter of degree. Low crosslinking density leads to slow recovery of foam and low compression set. Control, however, is difficult and crossing the line from slow resilience to compression set can be disastrous to a device.

The most effective way to control resilience is to practice the technology in U.S. patent 4,158,087 [18]. L. L. Wood, a pioneer in applications for hydrophilic polyurethane, discovered the effect that led to the polyurethane earplug industry. In his technique, a controlled amount of a polymer latex is added to the aqueous phase. When emulsified with a standard prepolymer, a slow-recovery foam results. In this application, the hydrophilicity of the polymer is of value in the processing of the part and is a detriment after processing. Therefore, the compelling reason for the product to be made from hydrophilic polyurethane is that the formulation must include significant amounts of the polymer latex. Table 8 is a summary of the data from the Wood patent.

Latex 1 was a soft styrene/butadiene latex (Latex LX430, MTP Kasei). Latex 2 was an acrylic (UCAR 874, Union Carbide). Latex 3 was a carboxylated styrene (67%)/butadiene (33%), the manufacturer was not noted.

The recovery test was to compress a 1-inch cube of the dry foam to 20% of its height and measure the time to recover.

TABLE 8. Low-Resilience Foam.

Example	Formulation	Amounts	Recovery Time (Seconds
1	Prepolymer Latex 1 Surfactant	80 g 80 g 4 g	16
2	Prepolymer Latex 2 Surfactant	80 g 80 g 4 g	15
3	Prepolymer Latex 3 Surfactant	80 g 80 g 4 g	2
Control	Prepolymer Water Surfactant	80 g 80 g 4 g	<0.6

As you can see the type of latex used has a dramatic effect on the recovery time. I don't know of a consistent theory as to how the latex affects recovery. The most commonly accepted explanation was offered by Mr. John Eagan [19], an important person in the development of hydrophilic polyurethane technology. He theorized that it was the adhesive nature of the soft latices that caused an adhesive effect.

3.9 DENSITY

We have discussed density as a way of controlling several of the properties above. Most important was the control of fluid capacity. We showed that the amount of fluid a foam will hold in its cells is one of the measures of hydrophilicity.

Control of density is achieved through several mechanisms. As always we can start with the prepolymer. The foaming process is, of course, caused by the evolution of CO_2. If a prepolymer is low in NCO (the source of the CO_2), then it follows that less gas is evolved and so higher densities will result (Table 9). An experimental prepolymer with which we have worked had a very low NCO and the resultant product was better described as a hydrogel than a foam.

Similarly, one can make high-density foams by putting an additive in the aqueous that does not release CO_2 when it reacts with the NCO. Among these compounds are alcohols and amines. W. L. Gore pioneered the pro-

TABLE 9. Void Volume
as a Function of Density.

Density	% Void Volume
5	92.8
6	91.3
7	89.9
8	88.5
9	87
10	85.5
11	84.1
12	82.7

duction of highly hydrophilic elastomers by adding difunctional amines with hydrophilic polyurethane prepolymers.

Adding methanol or ethanol to the aqueous has the effect of terminating the prepolymer chains. Each mole of alcohol prevents a mole of CO_2 from being released.

These techniques are commonly used, but their effects are typically too dramatic for most foam applications. When one wants to make minor changes in density to optimize the cost of manufacture, for instance, the processing conditions present the most practical solutions.

As we have said, and will discuss further in Chapter 4, once the aqueous and prepolymer come in contact with one another, two reactions occur, the production of CO_2 and polymerization. In a well-controlled process the evolution and the development of strength occur in such a way as to first trap the CO_2, converting its energy into expanding the mixture. Each of these reactions, however, has its own activation energy. Thus, if the temperature is changed both reactions change but at different relative rates. For instance, if the temperature is raised, the CO_2 reaction increases more than the polymerization. There is a danger that the internal energy generated by the expanding gas will overcome the strength of the polymer and burst through the surface to escape. The result is that the foam will expand and then collapse, producing a closed-cell, high-density mass. Similarly, but less dramatically, cooling the emulsion slows down the CO_2 reaction more than the polymerization and in the extreme also results in a closed-cell foam.

Catalysts are available to mitigate the effects described above, but typically one can control the process sufficiently to avoid the extremes and yet make small changes in density. In a well-controlled process, the operator is able to make changes in emulsion temperature on the order of 1 or 2 degrees and have a positive effect on the final product.

3.10 BIODEGRADABILITY

The final subject that we will cover in this chapter is biodegradability. There are incomplete and conflicting definitions of this important property. They vary from the definition used by many researchers in hydrophilic polyurethanes, suggesting a prejudice for the agricultural applications of this technology, that when buried, the material becomes indistinguishable from soil. At the other extreme, biodegradability is defined as the proportion of the mass that degrades to CO_2, water and biomass. Still further, and apropos medical devices, the most common disposal method for medical devices is incineration, an extreme form of biodegradability. Readers should consult a text on this subject to find the appropriate definition of their application.

The most common image of biodegradability involves digestion of the material by a microorganism. Although specific organisms have been identified that use polyurethanes as nutrients, for the most part, the hydrophilic polyurethanes that are commercially available are at best slow and don't pass the more stringent tests for degradation. To improve this it is necessary to include some biodegradable links in the backbone of the prepolymer or the foam.

The literature cites a wide variety of biodegradable linkages that have been used with polyurethanes. Among these are:

- starches
- sugars
- molasses
- lactic acid
- glycolic acid
- collagen

For the purposes of this book, U.S. patent 4,132,839 [20] serves as an example of how a prepolymer is made biodegradable. In an example the authors use to support their claims, polyethylene glycol 1000 is mixed with trimethylolpropane trilactate (351 grams and 60 grams, respectively) mixed with 225 grams of toluene diisocyanate. A catalyst was added continuously for 80 minutes at 60°C. The reaction was maintained for an additional 60 minutes at which time an additional 12 grams of TDI was added. The mixture was held at 60°C for another hour. A prepolymer of 25,000 cps. with an NCO of 2.4 meq/gm was recovered.

Foam was made from this prepolymer using more or less standard procedures and was then subjected to testing. In one experiment it was immersed in a buffer solution containing a proteolytic enzyme. The foam "completely degraded" in seven days at 25°C. A conventional hydrophilic polyurethane (without the lactate linkages) was unaffected by the enzyme treatment.

In another experiment, the foam was buried in compost for three months. Upon removal, the foam had started to fragment and could not be washed without falling apart.

It is assumed that the concentration of the trimethylolpropane trilactate will affect the rate at which the foam would degrade.

CHAPTER 4

Process Development

4.1 INTRODUCTION

In the first chapters, we have discussed the chemistry of the prepolymer and several case studies that convert the chemistry into commercial products. We then presented some techniques for how the chemistry and the physics of the foaming process are controlled in order to produce a product that fulfills the design requirements.

In the next chapters we will change directions slightly to focus on the more practical implications of manufacturing a device. In a sense, the earlier chapters are laboratory- or bench-level studies, while the following chapters will focus on "scale-up." We will discuss the common processes by which hydrophilic polyurethane parts are produced. This starts out with the chemistry of what happens when water is emulsified with the prepolymer, followed by a discussion of the stages of the reaction. The equipment used to manufacture foam is also discussed.

Subsequent chapters discuss the economics of foam production and the development of a quality system to control the safety and efficacy of the products.

4.2 THE CHEMISTRY

The isocyanate end groups on the prepolymer molecule are reactive to any compound with an active hydrogen. Thus, if a prepolymer is mixed with an alcohol or an amine, a reaction takes place that essentially caps the prepolymer and terminates the reaction.

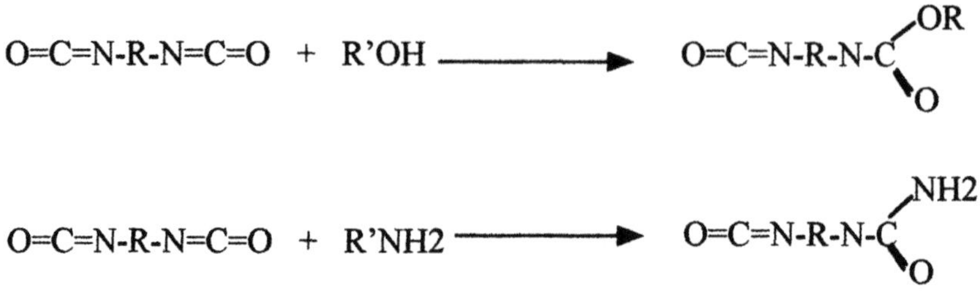

$$O=C=N-R-N=C=O \ + \ R'OH \ \longrightarrow \ O=C=N-R-N-C\begin{smallmatrix} OR \\ \\ O \end{smallmatrix}$$

$$O=C=N-R-N=C=O \ + \ R'NH2 \ \longrightarrow \ O=C=N-R-N-C\begin{smallmatrix} NH2 \\ \\ O \end{smallmatrix}$$

SCHEME 7. Reaction with amines.

You will notice that the product of the latter reaction is also an amine and therefore it follows that the amine product will continue to react with other isocyanate end groups. The result of this process is a continuous building of molecular weight until the isocyanate groups are consumed. This is the basis of the elastomer technology, with not only hydrophilic polyurethanes but all polyurethanes. This reaction is used as coatings for fabric, leathers and a host of other surfaces. When practiced with hydrophilic polyurethanes it produces a film with a very high water vapor transmission rate. When applied to a fabric as a continuous film, it is said to have high breathability. It has the typical disadvantages of hydrophilic polyurethanes in that it is physically weak and swells upon addition of moisture.

A variety of amines are used for this purpose. Each adds it own physical properties to the resultant film. The molecular weight, molecular structure and hydrophilicity of the amine contribute to the properties of the film. The reaction with amines is typically very fast relative to the reaction with water.

If the prepolymer has little or no crosslinking, the resultant elastomer can be thermoplastic. With crosslinking, however, the film can develop significant strength. If a chain-terminating component is added to the elastomer reaction, the molecular weight of the film can be limited to remain within the adhesive boundaries.

Thus by proper control of the reaction conditions, a high-moisture vapor transmissive adhesive film can be produced.

The reaction of primary interest, however, is the reaction of a hydrophilic prepolymer with water.

$$O=C=N-R-N=C=O \ + \ HOH \ \longrightarrow \ O=C=N-R-NH2 \ + \ CO_2$$

SCHEME 8. Reaction with water.

In this reaction, the production of CO_2 and an amine proceeds simultaneously to develop (ultimately) a stable foam. This is the core reaction of hydrophilic polyurethane foam technology.

The amine reaction, which we discussed above, develops the physical strength to contain the evolution of the CO_2. This would not take place, however, unless a radical change in the rheology of the reacting mass takes place early in the process. We will discuss this further when we describe the machinery on which these products are made, but it is important to consider the effect from a chemical perspective.

Once the water and the prepolymer are mixed, the rheology of the emulsion is that of a liquid. If it remained a liquid, the CO_2 would be able to escape the emulsion, and the result would be a closed-cell, high-density foam with little commercial interest. Commercially available prepolymers, as we have discussed, have a significant degree of crosslinking. Consequently, soon after reaction is initiated, the emulsion changes its rheology from a true liquid to a gel. It is part of the design requirements of a commercial prepolymer that there be sufficient crosslinking to rapidly develop enough gel strength to withstand the internal pressures developed by the evolving CO_2.

This is both a problem and an opportunity. Remembering that the evolution of CO_2 and the polymerization are two separate reactions, each with its own activation energy, it is clear that a change in temperature affects the rate of the reaction to a different degree. For instance, an increase in temperature of even a few degrees accelerates the CO_2 reaction more than it accelerates the amine reaction. In as much as it is the amine that produces the gelling of the mass, the CO_2 evolves at first in a liquid environment and, even if the emulsion has gelled, the CO_2 internal pressure may exceed the ability of the gel to contain it. The result is that the foam may expand initially, but the emulsion will reach a point where it will visibly collapse. This can be used to advantage if a high-density hydrophilic polyurethane is required, but typically this is not desired.

Alternatively, if the temperature is lowered, the strength of the gel increases faster than the rate of evolution of CO_2. The reader will note that the CO_2 reaction must take place first in order to produce the amine. Lowering the temperature has the effect of decreasing the difference in reaction rates. From a practical point of view, the gel strength develops so as to be able to withstand the internal pressures. To the observer, this is evidenced by a slower rate of rise, and the result is a higher-density product. In the extreme, a closed-cell foam is produced.

Thus it is clear that an efficient process will be able to juxtapose these reactions so as to produce the desired product. This is seen graphically in Figure 24.

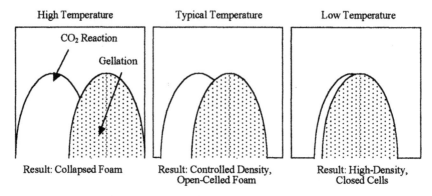

High Temperature   Typical Temperature   Low Temperature

Result: Collapsed Foam   Result: Controlled Density,   Result: High-Density,
                          Open-Celled Foam   Closed Cells

FIGURE 24. CO_2 and polymerization reactions.

It follows, therefore, that control of temperature is a critical process parameter. An efficient process will focus control efforts on the temperatures of the components, the degree to which the two phases are emulsified and other classical control methods. Once the emulsion is made and dispensed, there is little that can be done to control what happens. In this sense, the process changes to a more or less chaotic condition. From a control point of view, everything that can be done to moderate the process must be done before or during the emulsion stage. We will discuss some of the important control methods, but it is important to note that none of these methods is remedial in nature. We will also show, and emphasize in Chapter 7, that there are few corrective actions to recycle it. This has implication for the economics in as much as these two factors add up to a significant amount of scrap.

We will shortly turn our attention to the specifics of the processes by which these products are made but there is an important qualitative concept that must be discussed first. As we mentioned above two reactions take place once the water and the prepolymer are mixed. We will discuss this later, but it is an important engineering aspect that the reaction begins nearly instantaneously. If there are dead spots in the mix head, for instance, material will build up and react to plug off the equipment in a few minutes.

In summary, we have set the chemical stage for the manufacture of a device. We will now discuss the processes by which this basic chemistry is used to create products. Emphasis will be placed on control of the reactions. As we will describe, the control of the process is focused on the pre-emulsion phase. We have argued that once the water/prepolymer emulsion is made, little or nothing can be done to reaction mass to correct for noncompliances. Thus, we will focus the following descriptions on how the reactions are controlled.

The processes used to manufacture hydrophilic polyurethane foams are dependent on the product that is to be manufactured. A continuous roll of foam can be cast between release liners. This is considered to be the least expensive method. It is suitable for a device that can be diecut into the form in which it is used. Steel-ruled dies are commonly used for this purpose. We will discuss this later. Rotary dies, slitters, skivers and guillotines are also used to convert roll stock into useful devices.

Alternatively, buns can be made and converted through skiving and slitting and ultimately, diecutting. This method is lower in capital than a continuous line. Typically, the scrap rate is higher due to edge effects. Nevertheless, this is a convenient and common method of manufacture.

Molding is another common method for manufacturing devices from this material. We will discuss this later. The choice of materials for the molds and the use of mold release compounds are important aspects of molding devices.

While there are differences in the details of the above processes, they are all similar from a unit operations standpoint. Similarly, the equipment used has a common strain.

In this section, we will describe a typical process for manufacturing foam from several points of view. We will discuss the critical processing steps that are common to all of the processes described above. We will make the point that there are two stages of the process. In the first stage we focus all of our efforts on creating an emulsion of the prepolymer and an aqueous phase recognizing that the second stage can best be characterized as chaos. We will show that once the emulsion is made, we lose near complete control and the foam "does what it wants." Next we will describe the equipment typically used to finish the manufacture of the foam. Again, while each process varies in its details there is a good deal of commonality. We will continue by examining the stages that the emulsion goes through to produce an elastic, chemically stable foam. Lastly, we will discuss the common methods used to convert the raw foam into useful devices. This discussion will cover drying, slitting, skiving and diecutting.

4.3 CRITICAL PROCESSING STEPS

The flow diagram in Figure 25 is common to most if not all hydrophilic polyurethane foam manufacturing processes.

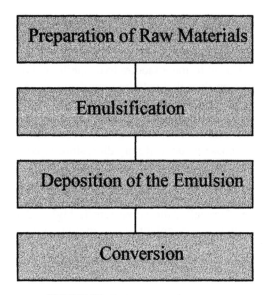

FIGURE 25. Critical processing steps.

4.3.1 Preparation of Raw Materials

The preparation refers to the treatment of the prepolymer and the aqueous phases before they are pumped into the emulsifier. This is typically tempering them with respect to temperature, and is usually done in a jacketed vessel.

The prepolymer tank is closed and is usually blanketed with dry nitrogen to prevent reaction with the humidity in the air. In as much as the prepolymer is a high-viscosity liquid at room temperature, it is not uncommon to see the prepolymer temperature raised to between 80 and 100°F. This lowers the viscosity enough to be pumped without fear of cavitating the pump. As a final control of the temperature, it is typically pumped through a heat exchanger large enough to ensure that the temperature of the prepolymer is controlled to within 1°F of a set point. If a temperature above room temperature is used, heated lines are recommended. Delivery of the prepolymer to the mixer is typically done using gear pumps. These are positive displacement devices that ensure a precisely controlled volume of material being delivered to the emulsifier. Care must be taken, however, to ensure that an uninterrupted flow of prepolymer to the pump (the low-pressure side) is maintained. Attempting to pump liquid at a rate faster than the prepolymer can flow into it results in cavitation, which changes the flow rates and can gel the prepolymer. It is important to remember that gear pumps don't suck material out of the tank. They just push material to its destination.

From a safety perspective it is important to minimize the chances of "dead heading" the pump (the high-pressure side). Gear pumps are capable of very high pressures. If a valve is closed between the pump and the emulsifier, the pump is capable of bursting even steel tubing. Standard engineering practice suggests including a recycle line with a pressure relief valve from the high-pressure side back to the low-pressure side.

Unfortunately, commercial pressure relief valves are mechanical in nature and have a tendency to become plugged with gelled prepolymer after a time, making them of little use. An effective method is to put a small piece of thin-walled plastic tubing in the line, which will serve as the weak link in the system. It has to be properly shielded, but in the event of a dead head, the thin-walled tubing would burst and prevent the build-up of the high pressure these pumps can generate.

Also shown in the diagram in Figure 26 is a continuous recycle line. Some equipment designers prefer to have the prepolymer (and the aqueous) flow through a valving system directly at or on the emulsifying mix head. The valves are arranged such that on demand the prepolymer (or aqueous) is directed from flowing through the recycle line to the mix head to be emulsified. The advantage is an instantaneous response, which is critical to a molding operation, which is typically batch in nature, but of less value to a continuous operation.

With regard to the construction materials, carbon steel is sufficient for the tanks. The preparation of the aqueous stream is similar. Due to the usually lower viscosity of the aqueous stream, a gear pump is not recommended. A progressive cavity pump (e.g., Moyno-type) is often used but not required. A consideration in the choice of pumps is the components of the aqueous. If it is to contain a solid (slurry or emulsion), a Moyno-

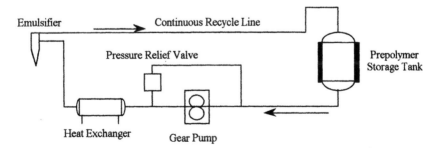

FIGURE 26. Prepolymer process flow.

type pump is preferred. When using latices, the shear forces created by the pump can coagulate the fluid.

Typically, the aqueous temperature is used to adjust foam quality. For instance, if a foam is found to have large cells and a low density, the temperature of the aqueous might be lowered. Thus, the aqueous temperature can be viewed as a primary means of fine-tuning a process. Other control methods exist (e.g., emulsifier speed), but aqueous temperature is a convenient method.

As with the prepolymer steam, the important parameters are temperature and flow rate. Any pump system that contributes to these will be effective. Pulsatory pumps (peristatic, piston, diaphram, etc.) should be avoided, but can be made to work if a pulse-reducing chamber or coil is used.

Depending on the nature of the aqueous stream, continuous agitation might be required. With that exception, a process flow scheme would be similar to that for the prepolymer.

In both cases, the prepolymer and the aqueous streams must be engineered in such a way as to safely deliver material to the emulsifier at a precise flow rate and a precise temperature. The absolute values are determined by a number of process-specific requirements, but the ratio of aqueous to prepolymer should be controlled to within 0.01 (e.g., 2.00 ± .01) and the temperatures should be controlled to within 0.5°F.

4.3.2 Emulsification

The next step in the critical path to manufacturing a hydrophilic polyurethane foam is the emulsification of the prepolymer and aqueous phases. The device that performs this operation is typically called the mix head. There is a wide variety of equipment whose purpose is to produce the emulsion. Some elaborate commercial mix heads have the capability to simultaneously emulsify as many as six separate streams. As we stated above, some mix heads are equipped with three-way valves so the streams can be continuously recycled until they are required to make foam.

Despite these engineering differentiations, the most successful designs fall into the category of what are called pin mixers. Figure 27 shows a typical design.

Although not shown, the mixer can contain stators. These are pins attached to the walls of the mix head; therefore, they do not spin. These stators increase the turbulence inside the mix head. As mentioned, some mix heads add up to six ingredients simultaneously. One mix head design has the capability to add a solid or powdered ingredient continuously.

The ingredients can come into the mix head on opposite sides. Several

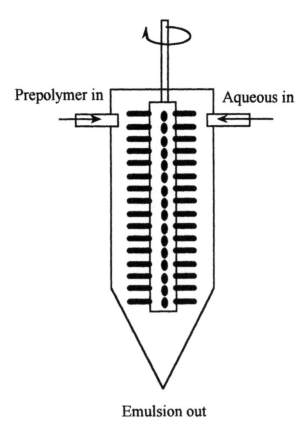

Emulsion out

FIGURE 27. A typical pin mixer for the emulsification.

successful schemes exist. Some bring the water and prepolymer in adjacent to one another. In others the two primary streams come in together. The only consideration seems to be the ability to clean out the mix head after a production session. In the above example, with the mixer spinning clockwise and the aqueous stream on the right, there can be a build-up of cured polyurethane on the leading edge of the prepolymer entrance site. With this exception, the relative position of the streams entering the mix head is not critical.

What results, typically, is a prepolymer in water emulsion. That is to say, the water is the continuous phase. Since this is essentially the last stage in which we can control the quality of the foam, there are a number of variables to be discussed.

As we mentioned above, the temperature is of critical importance in a well-controlled process. For the most part, it is the temperature of the emulsion leaving the mix head with which we should be most concerned. This is essentially determined by the temperatures of the component parts, but since a few degrees more or less can be significant, the mix head itself can have an effect on the emulsion. Depending on the speed at which the mixer spins, it can add 2–3°F to the emulsion.

It is the function of the mix head, however, to create a uniform emulsion. Not so much uniform in particle size, but with respect to time. It is the quality of the emulsion that, for the most part, defines the cell structure of the resultant foam. Emulsions of very small droplet sizes will result in a small-celled foam, all other things being equal. Emulsions with a broad droplet-size distribution will result in a wide variety of cells in the foam.

The surfactant is the primary determinant of the size of this emulsion, but the effect of the emulsifier cannot be ignored. The use of stators in the mix head increases the efficiency of the mixing process as does the speed of the mixer. Both of these factors tend to increase the temperature of the emulsion. When one seeks to control the quality of the emulsion with the mixer speed, which is commonly done, the temperature of the resultant emulsion is also changed, thereby complicating the control procedure. This then falls into the category of the art of foam production. Each time a change is made in mixer speed to address an emulsion problem, a corresponding change in the component temperatures is usually required. This, in turn, changes the emulsion conditions. Thus, a certain amount of operator intuition is required.

As we stated, it is the emulsifier that is the primary determinant of the size and distribution of the emulsion and the cell structure of the foam. Consult Chapter 3 for typical cell structures and surfactants. With regard to this chapter it is important to realize that it is the surfactant and the mixer that produce the emulsion. Once the emulsion leaves the mix head, the die is cast.

4.3.3 Deposition of the Emulsion

Once the emulsion leaves the mix head, it is placed into or onto a device that will define the form the foam will take. We have chosen a process to make continuous rolls as our primary focus, but we will begin by discussing some of the other forms the foam might take.

BUN STOCK

While the majority of hydrophilic polyurethane foam is produced on continuous rolls, the vast majority of all polyurethane foam is produced in buns. In large conventional (hydrophobic) polyurethane foam operations, the buns are the size of a small house. Focusing on hydrophilic polyurethane foams, however, the buns are produced at a size of 1 to 2 cubic feet. Part of the reason is the temperature sensitivity we discussed above. Since a few degrees' difference can affect the cells of the foam, one can imagine that the center of the foam will reach a different temperature than the edges. The center is virtually an adiabatic reactor. The temperature in the center can reach 140°F, while at the edges the foam will not exceed 85°F.

Another aspect of pouring big buns is the rate at which the molds are filled. If the time required to fill the mold extends past the time at which the system gels, flow patterns are created within the bun that are generally considered to be a negative. We will discuss this further when we discuss the sequence of reactions later in this chapter.

The last aspect of pouring buns is the drying of the product. With a typical hydrophilic polyurethane bun, it is necessary to skive the bun into thin sheets in order to efficiently dry the foam.

The use of mold release compounds is common in bun production. We will discuss this in more detail when we look at molded products. As the emulsion is poured into the tub, it is in its liquid state. As such it is capable of flowing into scratches and other imperfections in the tub, which results in a gradual build-up of material with repeated use. The choice of materials is important in minimizing this problem, but when making buns the cost of the tub is critical. This condition is somewhat mitigated by the use of compounds that minimize the adhesion of the curing polyurethane. These so-called mold release compounds vary from silicones to hydrocarbons and come in both water-based and solvent-based systems. There is no general rule as to which work best. For lab-scale investigations, I have found petroleum jelly to be useful.

Besides minimizing the adhesion of the foam to the tub, the mold release compounds can have a dramatic effect on the surface of the foam. By way of example, petroleum jelly produces a well-defined "skin" while lecithin, when used as a mold release, yields an essentially open-celled surface, i.e., little or no skin.

Despite the difficulties, making buns is a low-capital process and is capable of producing commercial-quality products at economical costs.

Molding

Molding is a subsection of bun making and many of the difficulties associated with making a bun also apply to molding. However, products that are molded are usually smaller than the typical bun. Also, the molding is typically of a more complex shape. This amplifies the need for mold release compounds.

Molding is separated into two categories, open and closed molds. The open-mold variety is the more common. This is also known as free-rise molding. In this process, the emulsion is poured into a cavity and is allowed to rise without restriction. This does not necessarily mean there is not a cap on the cavity. Quite often, after the emulsion is poured in the mold, a cap or top is placed over it. In the cap, however, vent holes are made to allow the air to escape. Typically, the mold is overfilled (with respect to the ultimate foam volume), so the vent also allows some of the curing emulsion to escape. The result is a molded part of low density.

Closed molds are used when the density of the product needs to be high. In this technique the mold is overfilled. It is then closed and clamped shut. The design of the mold and the time at which it is closed are such that air is allowed to escape but the foam is not. A compressed foam structure develops that results in a high density and very small cells. The strength of the device is characteristically high and this is usually the reason why the process is used. Great care must be taken, however. Extremely high pressures develop in the mold. Also, when the clamps are released, the mold tends to snap open with force.

The choice of mold material is important but not for the reason one might think. While we have worked with molds made from Teflon®, steel, aluminum, PVC, polyethylene, polypropylene, nickel-plated steel, chrome-plated steel and just about any other material imaginable, they all ultimately suffer from the same problem. Any imperfection will ultimately be the site of a build-up of foam and result in a nonreleasing or ripped part. This is true even with liberal use of release compounds.

The choice of mold materials depends on the complexity of the part to be made. For consumer-based products, or when cost is of primary importance, we have found that using semi-disposable thermoformed polypropylene molds is the best choice. With a suitable release compound, these devices can be used between 30 and 40 times. Their cost then permits them to be discarded and new ones used. These are typically placed in a form to give them the necessary stability.

When the part is complex and the cost permits, more rigid machined molds can be used. The choice of mold release compounds is of critical importance. You will find that after 30 or 40 uses, the mold will have to be removed from production and cleaned thoroughly. Hydroblasting has been used with some success.

There is a complexity with the molding process not associated with the molds themselves. It has to do with the batch nature of the process. This applies to the production of buns as well. We have pointed out that once the emulsion is created, it must be dispensed immediately. Imagine you have poured into a mold. Unless you, by interrupting the flow, begin pouring the next mold or bun, the emulsion will begin to cure in the mix head. Thus, any continuous process is preferred. A molding process is inherently batch, but it can be made to approach continuous by constructing the molding process so as to be able to pour molds without interrupting the foam. Anything less than a truly continuous process, however, will result in a gradual build-up of material in the mix head.

This situation is exaggerated in an inverse relationship to the size of the device. Under ideal conditions, one would want the part to be larger than the volume of the mix head. This has the effect of purging the mixer with each mold. For very small devices, however, this condition may not be able to be met. Thus, the flow of emulsion may be started and stopped several times without fully purging the mixer, which contributes to curing the emulsion and plugging it off. Purging the mix head between pours is often done, but the cost is substantial.

CONTINUOUS ROLLS

In a sense, this is also a bun or molded process. The molding is done in a sandwich of coated paper which is fed from long rolls. There is no difference in the mix head, but the emulsion is poured onto a continuously moving surface rather than a defined container. This surface is supported by some sort of table that can either simply serve as a support for the foam or be a mechanism to control the bottom surface temperature. For several reasons, a top surface is required for most products, so a coated paper is usually applied after the emulsion is poured. The diagram in Figure 28 shows a typical roll foam process using coated paper as the surface on which the emulsion is poured.

Typical of these processes is the use of silicone-coated release liners, a

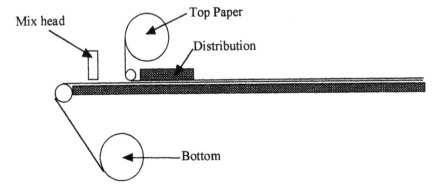

FIGURE 28. Continuous casting of foam.

variety of which are available. They vary in the amount of coating and the chemistry of the silicone. The bottom line is they have different release strengths with respect to the foam. A product development effort that involves roll stock should include an evaluation of which release liner should be used.

Other paper coatings are used. For example, polyethylene is common. It produces a more pronounced skin as evidenced by a more reflective surface.

An important characteristic of the process equipment is what is called the distribution box. As the emulsion leaves the mix head, it must be spread across the width of the table as quickly as possible. The manufacturers involved in the production of hydrophilic polyurethane foam have developed their own way of accomplishing this. The classical way of doing this is to oscillate the mix head across the width of the table and then lower the first roller to blend the ribbons of emulsion. This works well except for fast-curing foam formulations.

The physical movement of the emulsion must be completed before gelation occurs. Once spread, there is little to be done to the foam. Any control mechanisms and procedures will have to be applied before this. We will discuss this when we address the stages of reaction.

Once spread, the foam is allowed to rise and cure. This takes about 10 minutes, but the rate is strongly affected by temperature, the formulation and the definition of cure.

THE STAGES OF THE REACTION

We have discussed the change in the rheology of the emulsion from a true fluid to a gel. These are just two of several stages in the reaction that need to be considered in developing a process.

Cream Time

The time it takes for the emulsion to change from a liquid to a gel is called the cream time. The derivation of this term comes from pouring the emulsion into a cup. At a point in time, a foam rises to the top of the liquid and seems to separate as cream would from milk. Hence the term cream time. As we have discussed, this is the point at which the emulsion changes from a liquid to a gel. The processing significance is that if you are going to manipulate the emulsion in any way, it has to be done prior to cream time. Afterwards, any physical movement disrupts the very weak gel structure, leading to the release of CO_2 or causing nonuniformities in the structure.

Rise Time

As the gel develops its strength, it gradually becomes strong enough to withstand the expanding CO_2. At this point the physical volume stops increasing, thus setting the thickness of the foam. It is important to realize, however, that not all the CO_2 has evolved. Rather than increasing the volume of the foam, the internal pressure is exhausted through the foam and released to the atmosphere by bursting the windows on the cells. This is a necessary step in producing an open-celled foam.

Tack-Free Time

One of the principles of polymer adhesives is that if you can control the molecular weight of a polymer, you can control its adhesive properties. This is because as molecular weight develops, it goes through an adhesive phase. Typically, chain-terminating components are added to an adhesive formulation to limit the molecular weight to the adhesive zone. As the hydrophilic polyurethane foam develops molecular weight, it also passes through an adhesive zone. At the upper limit of that zone, it changes from an adhesive to a nonadhesive. This is called the tack-free time. At this point, it is strong enough to permit the removal of the top release liner. However, it is not strong enough to be handled. It is roughly 10% of its ultimate tensile strength.

This point has some potential uses to the product developer. As mentioned, there is a property of foam known as compression set. Any foam will be permanently deformed under compression if enough pressure and temperature is applied. Foams are rated with respect to how much pressure is needed to create the permanent deformation. Just after tack-free time, the compression set is extremely low, permitting a deformation with very little pressure. This point has been used as a convenient place to densify foam.

Cure Time

After the tack-free point, the strength gradually increases with any other notable stages. The cure time is usually defined as some function of ultimate strength. It is at this point that the foam might be strong enough to be handled.

Care must be taken, however. While the foam might have enough tensile strength to be handled, its compression set might still be low. If the material is immediately dried and rolled, it can take on a compression set produced by the rolling process. Rolling loosely minimizes this problem.

4.3.4 Conversion

The last category in the generalized process flow diagram is called conversion. In conventional polyurethanes, this discipline is usually limited to cutting the raw foam to size. For the purposes of this work, however, we have chosen to include the drying of the foam. Aqueous ratios between 1 and 2.5 are routinely used to manufacture hydrophilic polyurethane foams, which means that the cured foam will be between 50 and 70% moisture, which, typically, must be removed.

DRYING

Several types of dryers are used to remove the excess water, and their choice depends to some degree on the foam that is produced. The object of this operation is simply to remove water. It is beyond the scope of this book to recommend the best method. Rather, we would like to describe some of the ways people have used on a commercial scale to dry the foam.

Before we get into the details, it is important to describe the drying process from a semi-theoretical prospective. There are two diffusion processes at work. The first is the diffusion of the water vapor trapped in the void volume of the foam. This diffusion rate is relatively fast when compared to the diffusion of the water from inside the matrix material in the foam (Figure 29).

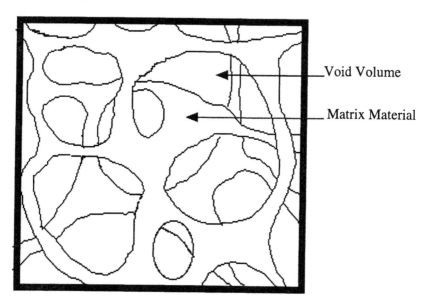

FIGURE 29. Foam structure.

Most of the water is trapped in the matrix of the foam, so the first step is to cause the water to diffuse into the void volume. Since the rate of diffusion is a function of temperature, any technique that heats this absorbed water faster will be a more efficient process. Once the water is vaporized, it must be flushed out of the voids. While this is relatively easy to do, the common drying techniques we will discuss vary in their efficiency to perform this function.

Suction Dryer

In this technique, hot air is sucked through the foam. The diagram in Figure 30 shows the basic scheme. The hot air will quickly heat the water in the matrix material and the scheme has the potential to flush out the water vapor in the void volume of the foam. For open-cell foams, this is an efficient drying scheme.

The box temperature can be as high as 250°F without fear of damaging the foam. The advantage of this equipment is that it is very fast. The rate of drying, however, is controlled by the temperature of the air and also the volume of air that can be pulled through the foam. We have discussed techniques in product design that seek to control the cell structure.

While the majority of foams we have discussed are of an open-cell structure, some foams are more open than others. Each foam will have its own permeability to air and when a dryer of this type is used, each foam will have its own drying curve. Foams with a tight cell structure will allow less air to permeate and therefore will dry more slowly than a more open cell structure. This property of foam is analytically measured in a test described in Chapter 6 of this book. It is a variation of the ASTM D-3574 method called air flow-through. This method describes a foam with respect to the volume of air that can be drawn through a foam at a given pressure drop or the pressure drop across a foam at a given flow rate through the foam.

It is clear that while this type of dryer is a fast and efficient way to dry open-structure foam, its use in more closed cell structures is problematic.

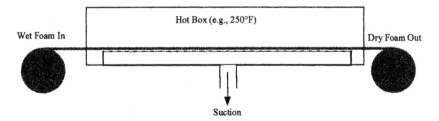

FIGURE 30. Suction dryer.

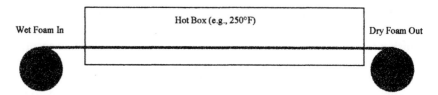

FIGURE 31. Convection dryer.

Convection Dryer

In this technique, the goal is to heat the water in the foam to vaporize it. The scheme is shown in Figure 31.

While the least expensive and probably the easiest to control, this technique suffers in its ability to heat the water in the center of the foam; also, it does not efficiently purge the foam of water vapor. Nevertheless, it is a common method. Its advantage over the suction dryer is that its drying efficiency is not a function of the cell structure.

A modification of the convection dryer scheme is known as an impingement convection dryer. In this technique, a stream of air is directed on the surface of the foam. While useful with some materials, it does not significantly help drying these types of foam.

Microwave/Radio-Frequency Drying

Hydrophilic polyurethane foams can also be dried by this technique. The advantage is that the water in the matrix material is quickly heated to vaporize it. The volume of water vapor itself serves to flush all but a small fraction of the water remaining.

The disadvantages of this technique rest solely in the cost to purchase, install and operate the device.

Tumble Dryers

The last technique is typically used for the more batch-type processes, while the other drying equipment we have discussed is suitable for continuous operations. Tumble dryers are best described as commercial-scale clothes dryers. That is, they are convection devices with the advantage of being insensitive to foam structure, but they are slow relative to the other techniques.

Tumble dryers have another disadvantage. If the product is not fully cured, the parts can stick together. This can be mitigated by using talc, cornstarch or fumed silica, but when that is not possible, ensuring that the parts are fully cured by extending the curing time or preheating the product will be necessary.

There is one more aspect of drying that must be considered. In the techniques described above, the temperature of the foam reaches the boiling point of water at the end of the drying cycle. If the foam is heated past the removal of the water, its temperature will increase above that point. Hydrophilic polyurethane foam, as with all aromatic polyurethanes, has a tendency to yellow quickly at elevated temperatures. The graph in Figure 32 is typical of this type of foam (consult Chapter 6, Section 6.11, item (4)*a*.

Care should be taken, therefore, not to overdry the foam. A typical specification will list the % moisture as between 0 and 5%, allowing some manufacturing latitude.

DIECUTTING, SLITTING AND SKIVING

The last topic in our survey of hydrophilic polyurethane process development is the conversion of the raw, dry foam into the final form. This discussion deals specifically with products made from buns and roll stock. It is assumed that molded products leave the process in the final form and therefore need little or no work before shipment. The only exception is the possible trimming of flashing, which results from foam material seeping into the seams of a closed mold.

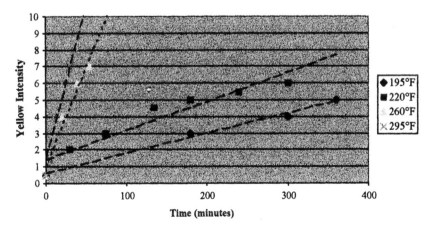

FIGURE 32. Yellowing of foam by temperature.

With buns and roll stock, however, the material typically must be processed to achieve its final, commercial form. This processing can be as simple as slitting the edges of the foam to remove the rough edges or slitting to a width that is appropriate for a subsequent piece of equipment. Many times the foam is laminated to another material. Slitting is used, for instance, to make ribbons of foam, 4 inches wide. The foam is subsequently loaded onto a packaging machine that guillotines the ribbon to length and automatically seals it into a package. Some wound dressing manufacturers use this technique. Slitting is also common when the foam is to be cut in a rotary diecut operation.

A number of different designs are used for slitters. In our experience, what works best is a slitter that uses a shearing action with a rotating blade the edge of which is pressed against a flat stationary surface.

Buns produce foams that are typically thicker than needed. In this case an operation called skiving is performed, which slices a thick bun into thinner sheets. This operation is performed using very fine tooth saws. Thicknesses as small as 0.1″ are possible by this technique.

Lastly, and most commonly, the material is diecut to shape. There are two common techniques, steel-ruled dies and rotary dies. The former is more common in this business. A thin blade is mounted in a dieboard to support it, and with the help of a hydraulic press enough pressure is exerted to cut the foam. Dieboards can be constructed into very complex sheets.

It is important to note that the action of the blade on the foam is a crushing action. The press exerts vertical pressure on the foam to cut it. This has a number of implications, not the least of which is the compression set phenomenon we discussed earlier. With freshly made foam or foam designed to have a low compression set, using a steel-ruled die permanently deforms the edges of the part. This can be used to advantage in as much as it can produce an attractive, rolled-over appearance. When this is not wanted, a guillotine type of cutter is used. A guillotine uses a shearing action to cut the foam.

There are a number of types of blades. Apart from the materials from which they are made, the position and the angle of the blade (the bevel) are important characteristics. In our experience an inside bevel of 60° is of general interest.

CHAPTER 5

Economics

5.1 INTRODUCTION

We have emphasized that in order for a hydrophilic polyurethane project to be successful, there has to be a compelling reason for it to be hydrophilic or water-compatible. The reason is economic. We will show in this chapter a method to calculate the cost of hydrophilic foam. Although the primary purpose of this discussion is to serve as a guide in determining the cost/value of a product during the formative stages of a project, it is often useful to know if the cost is going to permit commercialization. If the cost is too high, the project can be terminated. Alternatively, an early economic evaluation guided by the procedures in this chapter, might refocus a research project to cost cutting.

A classic project in this regard was the development of a dressing for a tracheal tube. A preliminary investigation showed that the patient acceptance would be based on the fact that dressings made from hydrophilic polyurethane foam were very soft and comfortable. The absorbency of the dressing was nice but it wasn't critical. The project proceeded well from the technical perspective, but upon the realization that the cost would be $0.12–0.15 per dressing, the project was killed because the alternative (a conventional polyurethane foam) cost $0.02–0.03.

Although we have attempted to cover all of the factors that contribute to the manufacturing cost, each project is different. Small design characteristics can and often do have a dramatic effect on cost. A round dressing is much more expensive to produce than a square dressing, for instance. This is due to the amount of waste produced during diecutting.

We will begin with the costs of the components of a foam and supplement this with labor and indirect costs. We will then discuss the effect of scrap on the net cost of manufacture. We conclude with a consideration of the sensitivities in costs.

5.2 COST OF PREPOLYMER MANUFACTURE

We begin our discussion with the cost of production of the prepolymer. If it is the intention to purchase hydrophilic polyurethane prepolymer from a supplier, one can enter the expected price ($3.00 U.S. is a good number). If it is within the scope of the project to make the prepolymer yourself, or if you just want to maintain control of the overall economics, this discussion is appropriate.

Whether the prepolymer is made by a batch or a continuous process, the prepolymers that are currently available, and of interest to us, are based on the reaction of a polyol and a diisocyanate. We discussed this in detail in Chapter 4. We also noted that the reaction might include a crosslinking agent. It is important to know that there are no recoverable by-products of this reaction. For all intents and purposes, the reaction can be assumed to go to completion without by-products. It takes two molecules of toluene diisocyanate (MW = 174) to react with one molecule of PEG 1000 (1000 MW polyethylene glycol). Thus, 1000 grams of PEG requires 358 grams of TDI. It is typical, and many people think it is required, that the prepolymer requires an excess of TDI. There are also dimer and trimerization reactions, as well as amine/isocyanate reactions. Remembering that all of the by-products go into the final prepolymer, the actual ratio of polyol to isocyanate is typically around 2:1. For the purposes of this evaluation we will use this ratio. We will also assume that the manufacturer will use a chemical crosslinker to the extent of 3% of the prepolymer.

Table 10 describes the raw material cost of the prepolymer. This analysis produces 920 pounds of prepolymer per hour.

The labor to produce the material is described by the next table along with a summary. In Table 11, we are adding a manufacturing overhead column. This represents the cost of capital, heating, employee benefits and

TABLE 10. Direct Raw Material Costs.

Component	$/lb.	lbs./hr.	$/hr.
TDI	$1.50	300	$ 450
PEG 1000	$1.30	600	$ 780
Crosslinking	$3.20	20	$ 64
		920	$1,294
Raw material cost ($/lb.)			$1.41

TABLE 11. Prepolymer Labor and Overhead.

Position	$/hr.	Manufacturing Overhead	Total Labor
Operator	$15.00	$22.50	$37.50
Operator	$15.00	$22.50	$37.50
Supervisor	$25.00	$37.50	$62.50
Production rate			920 lbs./hr.
Labor rate			$0.11/hr.

other overheads. The manufacturing overhead is assumed to be 150% of the cost of labor.

The cost of manufacture is the raw material cost plus the labor cost. Adding the sales and administrative costs justifies a cost for sale in the range of $3–3.25/lb.

If the above prepolymer is to be used in a medical device, we will show that an evaluation of the prepolymer cost also represents the raw material cost of the medical-grade foam. This is because the aqueous, the other component of the foam reaction, is typically very inexpensive relative to the prepolymer. Once dried, the mass of the foam is essentially the mass of the prepolymer.

5.3 AQUEOUS

For a typical medical-grade hydrophilic polyurethane foam, the aqueous contains nothing more than an emulsifying agent. The concentration is typically less than 0.5%, so the raw material cost of the aqueous is very low relative to the prepolymer. To make the economics more interesting, we will build them with an aqueous that includes an "active" ingredient. If the device to be produced is medical, the component might be an antibiotic or antimicrobial. Similarly, a cosmetic foam might include a fragrance or a soap. In any case, the aqueous with an expensive additive will have a value and a cost. In Table 12 we examine a system that includes an emulsifier and an active ingredient. Water is assumed to be free. This is typically not the case.

5.4 FOAM PRODUCTION

We must now use these raw materials to make a foam. In order to calculate the costs associated with the manufacturing process, we must describe the process. We will again use the process model introduced in

TABLE 12. Manufacturing Cost of an Aqueous.

Component	$/lb.	%	Cost/lb.
Water	$ 0.00	89%	$0.00
Emulsifier	$ 3.00	1%	$0.03
Active ingredient	$10.00	10%	$1.00
			$1.03
Aqueous labor ($/lb.)		$0.01	
Aqueous cost ($/lb.)		$1.04	

Chapter 4. Refer to Figure 25 in that chapter, but for review we identified several unit operations.

- prepolymer production
- aqueous preparation
- foam production
- foam drying
- conversion

We have discussed and quantified the first two operations, but in order to describe the next categories, we will have to assign some values that represent size or output of the process. Although we will be using the continuous process model, many of the rationales we use are also applicable to other processes. Table 13 gives us the process conditions under which the calculation will be made.

These capacities are controlled by several factors. The rate-determining step might be the length of the line. If it is determined that the foam takes 10 minutes to cure sufficiently to be handled, and the foam line can only be 100 feet long, the fastest the line can operate is 10 feet/minute. At 24 inches wide, that translates to 1200 sq. ft./hr. Alternatively, the

TABLE 13. Foam Production
Plant Capacities.

Web width (in.)	24
Foam density (lbs./cu ft.)	6
Thickness (in.)	0.25
Production rate (sq. ft./hr.)	1200
Production rate (cu.ft./hr.)	25
Production rate (lbs./hr.)	150

process might be dryer limited. Depending on the types of products produced in a facility, the rate-determining operation may be different for each product.

For this analysis we will assume the plant is limited to 1200 sq. ft./hr. due to the length of the foam line (not including the dryer) and the cure time of the foam.

Using these data, Table 14 represents the direct raw material costs for a process that uses a 1:1 aqueous to prepolymer ratio.

You will remember that the aqueous contains some nonvolatile components (the emulsifier and the active ingredient). In determining the flow rates of prepolymer and aqueous, the mass of the resultant foam is the sum of the prepolymer and the non-volatiles.

Other materials are required to manufacture the foam, however. Among these are boxes, bags and other packaging materials. If the process includes casting the foam between a sandwich of silicone-coated release paper, the packaging costs can be considered insignificant, at least for this analysis.

Another cost that must be considered is the cost to dry the foam (assuming it has to be dried). For this analysis we will use the figure $10 per million BTU. Table 15 summarizes these indirect costs.

The last category to be covered in the production of the foam is the labor costs. We will use the manufacturing overhead procedures we used with the prepolymer. For this analysis we will assume the process requires the work of four operators and a quality control technicians and a supervisor. The hourly rate is listed in Table 16.

We have now described all of the factors that go into the manufacture of a roll of hydrophilic polyurethane foam including drying. To summarize, review Table 17.

TABLE 14. *Direct Raw Material Costs of the Foam Process.*

Component	Lbs./Hr.	Cost/Hr.	Cost/Hr.
Prepolymer	135.1	1.51	204.00
Aqueous	135.1	1.04	140.50
			344.51
		Rate	Cost/Unit
Production rate (lbs./hr.)		150	$ 2.30
Production rate (cu. ft./hr.)		25	$13.78
Production rate (sq. ft./hr.)		1200	$ 0.29

TABLE 15. Summary of Indirect Production Costs.

Indirect Cost: Energy

Wet Foam (lbs./hr.)	Dry Foam (lbs./hr.)	Water (lbs./hr.)	Cost/Hr.
270.2	150	120.2	$1.41

Indirect Cost: Release Paper

Dry Foam (sq. ft./hr.)	Wet Foam (sq. ft./hr.)	Paper (sq. ft./hr.)	Cost/Sq. Ft.	Cost/Hr.
1200	1800	3600	$0.02	$72

		Cost/Unit
Production rate (lbs./hr.)	150	$0.49
Production rate (cu. ft./hr.)	25	$2.94
Production rate (sq. ft./hr.)	1200	$0.06

TABLE 16. Labor Cost of Foam Production.

Position	Hourly Rate	Manuf. Overhead	Cost/Hr.
Operator	$10.00	$15.00	$ 25.00
Operator	$10.00	$15.00	$ 25.00
Operator	$10.00	$15.00	$ 25.00
Operator	$10.00	$15.00	$ 25.00
QC tech.	$15.00	$22.50	$ 37.50
Supervisor	$25.00	$37.50	$ 62.50
			$200.00

		Cost/Unit
Production rate (lbs./hr.)	150	$1.33
Production rate (cu. ft./hr.)	25	$8.00
Production rate (cu. ft./hr.)	1200	$0.17

TABLE 17. Summary of the Manufacturing, Cost of Roll Stock.

Units	Direct Material Cost	Labor Cost	Other	Total
Pounds	$ 2.31	$1.33	$0.49	$ 4.13
Cubic feet	$13.83	$8.00	$2.94	$24.77
Square feet	$ 0.29	$0.17	$0.06	$ 0.52

5.5 EFFECT OF DENSITY ON COST

TABLE 18. Process Description
at 2 Lbs./Cu.Ft.

Web width (in.)	24
Foam density (lbs./cu ft.)	2
Thickness (in.)	0.25
Production rate (sq. ft./hr.)	1200
Production rate (cu. ft./hr.)	25
Production rate (lbs./hr.)	50

You will remember that in the description of the process, we assumed that the density of the product would be about 6 pounds per cubic feet. I want to restate the above graph at another density to make a point. Thus at 2 lbs./cu. ft. the summary above is restated in Tables 18–20.

Since the other costs in the process are not affected by the density, Table 21 summarizes Table 10, the direct raw material costs.

As you can see, the per square foot cost is reduced to $0.33 from $0.52, a 36% decrease. This is the primary reason why there needs to be a compelling reason for the product to be hydrophilic. While hydrophilic polyurethane foams can have densities as low as 5 lbs./cu. ft., values less than that are not possible. The reason is that in order to make usable foam at lower densities, the foams need to be relatively strong. Strength, as we have discussed, is not a characteristic of hydrophilic foams. Conventional polyurethanes, however, are strong enough to be made at densities as low as 1.5 lbs./cu. ft. This gives them an advantage that cannot be overcome.

Looking at it another way, this gives us the possibility of gauging the relative value of the products. The conventional polyurethanes, on one

TABLE 19. Summary of the Direct Raw Materials.

Component	Lbs./Hr.	Cost/Lb.	Cost/Hr.
Prepolymer	45	$1.52	$ 68.40
Aqueous	45	$1.04	$ 46.80
			$115.20
	Rate	Cost/Unit	
Production rate (lbs./hr.)	50	$2.30	
Production rate (cu. ft./hr.)	25	$4.61	
Production rate (sq. ft./hr.)	1200	$0.10	

TABLE 20. Summary of Indirect Materials.

Indirect Cost: Energy			
Wet Foam (lbs./hr.)	Dry Foam (lbs./hr.)	Water (lbs./hr.)	Cost/Hr.
90	49	41	$0.48
Indirect Cost: Release Paper				
Dry Foam (sq. ft./hr.)	Wet Foam (sq. ft./hr.)	Paper (sq. ft./hr.)	Cost/Sq. Ft.	Cost/Hr.
1200	1800	3600	40.02	$72

		Cost/Unit
Production rate (lbs./hr.)	150	$0.49
Production rate (cu. ft./hr.)	25	$2.94
Production rate (sq. ft./hr.)	1200	$0.06

hand, are used as filters or packing material, whereas hydrophilic poly-urethane foams have a functional purpose with their ability to absorb aqueous-based fluids. Thus, all things being equal, the ability of the hydro-philic foam warrants the 57% premium in cost (33% to 52%).

5.6 SCRAP

There is another important factor that must be considered. As we discussed in Chapter 4, despite our attempts to control everything from temperatures to formulation, once the emulsion leaves the mix head, the process is chaotic and for the most part out of our control. Because of this and other factors that we will discuss, a significant amount of scrap is produced.

TABLE 21. Summary of the Costs of a 2 Pound Density Foam.

Units	Direct Material Cost	Labor Cost	Other	Total
Pounds	$2.30	$1.33	$0.49	$ 4.12
Cubic feet	$4.61	$8.00	$2.94	$15.55
Square feet	$0.10	$0.17	$0.06	$ 0.33

Scrap falls into several categories. Some of these categories are avoidable or can at least be minimized. We will discuss each briefly and describe their typical magnitude. We will then summarize and show their effect on the manufacturing cost.

5.6.1 Start-Up Scrap

Again, using the model of a continuous process, once the emulsion is produced in the mix head, it is dispensed on a table, typically between sheets of release paper. We have discussed the sensitivity of the prepolymer reaction with water to temperature. Even though the aqueous and prepolymer are tempered, the rest of the equipment will need to come up to temperature. This may take a few minutes. Further, the table on which the foam will be poured needs to be heated (typically) to some equilibrium value, and this may take as long as 0.5–1 hour depending on the temperature of the emulsion.

If the foam has some peculiarity, this start-up period can be even longer. If, for instance, the cell size needs to be held within tight limits, this can greatly extend the start-up time.

5.6.2 Edge Scrap

If the foam to be produced is in the form of rolls, the edge of the foam is usually not of acceptable quality. Typically, surface effects will produce a falling off at the edge as opposed to a square edge. This is usually described in the specification for the foam as the actual thickness versus the usable thickness. The way the price is developed usually establishes the procedures for how to handle this. The client should not, however, be charged for materials that cannot be used, so this leads to a certain amount of scrap.

5.6.3 Out-of-Spec Product

Despite our best efforts, it is the nature of all manufacturing processes to produce material that is noncompliant. During the development stages of a project, the manufacturer should establish the capability of the process vis-à-vis the precision of the process. For example, typically the designer will require foam of a specific thickness. By the time the project reaches commercialization, the manufacturer should know how well he can control this parameter. Statistical control charts have been developed that describe the process in terms of upper and lower control limits. These are defined as 3 standard deviations (3 SD) above and below the average, respectively. Also, as part of the definition, these upper and lower limits represent 99.73% of the production.

If the client and customer are satisfied with the level of precision described by the 3 SD, the specification will reflect that. It is rare, however, that this is the case for hydrophilic foam. The specifications typically include tolerances that are within the 3 standard deviations. This imposes a certain level of scrap that must be accounted for in the net manufacturing cost. Parenthetically, it creates an environment that gives the manufacturer an incentive to improve the process.

5.6.4 Conversion Scrap

If the product of the process is roll stock, the above categories describe the scrap rate. If, however, the manufacturer is required to convert the product in some way, additional scrap is produced.

If, for instance, the manufacturer is required to trim off the edges of the rolls (called "slitting"), the edge scrap needs to be considered. Just the fact that this is a separate operation, and requires handling, will create a scrap rate.

If the material is to be sliced, additional scrap is created. This is most common with manufacturers who make buns. Once produced, buns need to be sliced into sheets (called "skiving").

Diecutting presents a variety of problems with respect to scrap. Consider the diecutting of circles from a sheet of foam. In this example (Figure 33), clearly, most of the foam could be considered scrap. Even if the

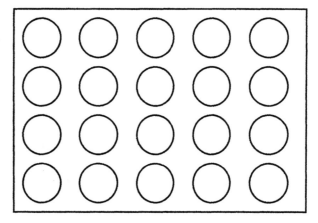

FIGURE 33. Diecutting circles from a web.

circles were packed as closely as possible, a significant amount of scrap would be produced. Thus, the process of diecutting causes scrap, which needs to be considered in considering the cost of manufacture.

Although they are not part of the model process we are surveying, a word needs to be said about molded products. On a pound-for-pound basis, it is generally true that a molded product produces more scrap that the cast products we are examining. This is because the molding process leads to a significant category of scrap not usually prevalent in cast materials. This category involves incomplete filling of the molds or trapping of air. This in turn leads to parts that can have significant structural defects causing them to be rejected. This has another disadvantage, each part must be inspected, further adding to the cost.

5.6.5 Summary of Scrap

All of these categories add up to a significant amount of rejected material. It is the obligation of the manufacturer to evolve processes that minimize this, but rejects are a fact of life. Thus, in as much as all members of the team, designer/marketer and manufacturer, need to make a profit from the venture, increases in cost must be passed along to the consumer. A successful product is by definition a product that can be priced such that each member of the team makes a decent return.

Each project will have a characteristic scrap rate, which there is no way to generalize. In order to illustrate the effect on cost, it is most convenient to show the effect of various scrap rates on the cost of manufacture. Table 22 calculates the net cost of manufacture at various scrap rates. The calculation is a simple proportionality. If the scrap rate is 50%, the cost of manufacture doubles. We will take the original example (the 6-pound density foam).

TABLE 22. Effect of Scrap Rate on the Cost of Manufacture.

Scrap Rate (%)	Net Cost of Manufacture ($/sq. ft.)
0%	$0.52
10%	$0.58
20%	$0.65
30%	$0.74
40%	$0.87
50%	$1.04

5.7 SENSITIVITIES

In Table 22, we used a technique called sensitivities. The purpose of presenting data in that manner is a recognition of the fact that the analysis described in this chapter is a guide, not an attempt to define the cost. This can only be done after the development process. In effect, Table 22 shows how sensitive the cost of manufacture is to the scrap rate. In the development of the model in this chapter, we assumed a number of factors that in real life are, in part, subject to the development process. As an integral part of the development process, the manufacturer has the responsibility to define the economics with as few assumptions as possible. What we have discussed here is a preliminary estimate.

For this reason it is helpful and realistic to come up with a series of sensitivity graphs that help develop a range of costs that might be experienced. As we stated in the beginning of the chapter, this might have the effect of focusing the research effort on cost cutting or, alternatively, kill the project outright.

When the principles taught in this chapter are applied to a project, rather than deriving a single cost number, the result should be a series of graphs that describe the most important cost factors and the likelihood of financial success.

CHAPTER 6

Analysis of Hydrophilic Polyurethane Foams

We have been using a paradigm to describe the process of developing a device or product from hydrophilic polyurethane. It began with the description of a device by a client supposedly familiar not only with how a product should function, but also with the necessary differentiation to make it a commercial success. That definition was then converted to a device by combining the chemistry and the physics of the hydrophilic polyurethane coupled with certain process techniques and procedures. We will be discussing the economics and describing how the elements of the development process are combined into a quality system.

In this chapter, we will talk about how the product is evaluated with respect to the design, i.e., does it fulfill the original expectations? Also, the foam or the device from which it is made need a series of defining characteristics to ensure that it is of a consistent quality once its manufacture is expanded to commercial size. In effect, a set of in-process tests will have to be developed that ensure product quality. These characteristics will have to be built into a specification document.

Lastly, in the area of medical devices, it is necessary to be able to prove the efficacy of the device. Still further, the owner of the device will have to assure the FDA that the product is made and tested in such a way as to guarantee uniformity, lot to lot.

For these reasons, and to avoid the complications that arise from a manufacturer and owner having different definitions for a given property, we have developed a series of analytical methods that we have found to be useful in characterizing hydrophilic polyurethane foam and the devices made from it. The American Society for Testing of Materials (ASTM) has the largest inventory of approved methods of analysis and testing. Unfortunately, none is specifically intended for use with hydrophilic foam. One test, however, comes close. Its reference number is D3574.2. It is a set of procedures intended to be used to evaluate the physical properties of a multicellular material.

The methods, however, do not cover the absorption characteristics of a hydrophilic polyurethane. Nor do they cover, specifically, wet foams. Thus the procedures need to be modified in order to fully describe the device that was the object of the design exercise.

While it is advised that manufacturers and product designers use ASTM D-3574 or close modifications, many labs do not have the proper analytical equipment. For this reason we have developed methods that utilize common equipment in the typical lab. Specifically many of the tests that are included in D-3574 are approximated by the use of an electronic balance and a device that permits the gradual compression of the foam. This compression device is described in Figure 34.

A sample of foam (4) is placed directly on the balance (5) and the compression device is positioned over it. The screw (3) lowers the plate (1) until it just touches the top of the foam. The thickness of the foam is read off the micrometer dial (6). Methods will follow that show how this device is used to characterize many important characteristics.

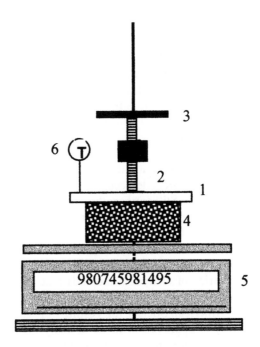

FIGURE 34. Compression device.

While D-3574 should be used whenever possible, not all development labs have the specialized equipment required. For this reason, we have modified the D-3574 test so an estimate of the tensile properties can be determined with an electronic balance. While the compression tests produce values that are reasonably close to those that would be derived from the D-3574 method, they must be used with caution. The reason for this has to do with the rate at which the samples are pulled. As we will describe in the method, this is critical to the tensile strength at break, for instance. In other words, one would get a different value at break for a sample that is pulled quickly versus one that is pulled very slowly.

Nevertheless, if the values derived by this method are compared with samples pulled at roughly the same rate, some valuable information can be extracted.

The tensile properties can be estimated by utilizing the hook on the bottom of many electronic balances. While it is intended to be used to determine the density of samples (Archimedes' method), it is useful for our purposes. A bracket must be fabricated that holds a piece of foam. It must also provide for a wire to hang it from the hook on the balance. Finally, a bracket must be fabricated to hold the foam at the other end (see Figure 35.).

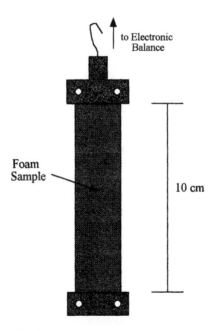

FIGURE 35. Brackets for tensile properties.

After being hung from the bottom of a balance, the bottom bracket is slowly pulled down and the reading on the balance monitored. At the point it breaks, the force is read and the elongation noted on the appropriate scale. Thus the tensile strength at break and the elongation are estimated.

Some tests included in the ASTM set do not need alteration and in these cases the ASTM method should be used. The compression set and rebound tests are examples.

With that as an introduction, we will now give the details of typical tests used to evaluate hydrophilic polyurethane foam. The first of those procedures is a technique to hydrate a foam to prepare it for an analysis of its wet properties.

6.1 PRE-HYDRATING A HYDROPHILIC FOAM FOR SUBSEQUENT ANALYSIS

(1) Scope
   a. This test method covers pre-hydrating of a foam in preparation for the determination of the wet methods that will follow.
(2) Test specimen
   a. A representative specimen of not less than 10 cc in volume shall be cut from a portion free of voids and defects. When these dimensions are not possible, a representative portion is agreed upon.
(3) Number of specimens
   a. One specimen shall be tested.
(4) Procedure
   a. Place the specimen in water at room temperature.
   b. Squeeze and release the foam under the water to remove entrained air.
   c. Leave the foam in the water for a minimum of 30 minutes, or until it is ready for subsequent testing.
   d. Remove the specimen from the water and squeeze it to remove the water. Use paper towels to pat the foam dry.
   e. Analyze immediately. If there is a delay in analysis, repeat the hydration procedure.

6.2 FOAM DENSITY

(1) Scope
   a. This test method covers determination of the density of foam by calculation from the mass and volume of the specimen. The den-

sity value applies to the immediate area from which the specimen has been taken. It does not necessarily relate to the bulk density of the entire roll or pad.

(2) Test specimen
  a. Interior density—a representative specimen of not less than 10 cc in volume shall be cut from a portion free of voids and defects and as near as possible to the section from which the tension and tear specimen will be taken if those tests are to be performed. When these dimensions are not possible, a representative portion as agreed upon by the seller and the purchaser shall be used.

(3) Number of specimens
  a. One specimen shall be tested.

(4) Procedure
  a. Determine the mass of specimen in grams to within 10 milligrams [two (2) decimal places].
  b. Determine the dimensions of the specimen in centimeters and calculate the volume. Care should be taken in measuring the dimensions of the specimen. In no case should a pressure greater than 70 Pa be exerted on the specimen when it is measured.

(5) Calculation
  a. Calculate the density in pounds per cubic foot as follows:

$$\text{Density} = M/V$$

  Where

  M = mass of specimen in lbs. = mass in grams/454
  V = volume of specimen cubic feet = volume in cubic centimeters/28317

(6) Report
  a. Determine the density to the nearest 0.1 lb./cu. ft.

6.3 INDENTATION FORCE DEFLECTION AT A SPECIFIED DEFLECTION

(1) Scope
  a. This will be known as the indentation force deflection test and the results as the IFD values. This test measures the force necessary to produce 25 and 65% or other designated indentations in the foam product.

(2) Apparatus

    a. An apparatus shall be used having a flat circular indentor foot 323 cm^2 in area connected by means of a swivel joint capable of accommodating the angle of the sample to a force-measuring device. The apparatus shall be arranged to support the specimen on a level horizontal plate.

(3) Test specimen

    a. The test specimen shall consist of a representative portion of the sample, free of visible defects, except that in no case shall the specimen have dimensions less than 380 by 380 by 20 mm. Specimens less than 20 mm thick shall be plied up without the use of cement, to a minimum of 20 mm.

    b. The IFD values for molded products are dependent on the specimen dimensions. Higher values are generally obtained for specimens that contain all molded surfaces.

(4) Number of specimens

    a. One specimen shall be tested.

(5) Procedure

    a. Place the test specimen in position on the electronic balance. The specimen position shall be such that whenever practicable the indentation will be made at the center of all articles, except where another location is agreed upon by the parties.

    b. Preflex the area to be tested by twice lowering the indentor foot to a total deflection of 75 to 80% of the full-part thickness. Allow the specimen to rest 6 ± 1 min after the preflex.

    c. Bring the indentor foot into contact with the specimen and determine the thickness. Indent the specimen to 25% of this thickness and observe the force in grams after 60 ± 3 s. Without removing the specimen, increase the deflection to 65%, allowing the force to drift while maintaining the 65% deflection, and observe the force in grams after 60 ± 3 s.

(6) Calculation

    a. Calculate the 25% and 65% IFD in pounds/square inch as follows:

$$\text{IFD} = \text{force/area of the foot}$$

where:

$$\text{Force} = \text{force in grams/454}$$
$$\text{Area} = \text{area of the foot in square inches/144}$$
$$\text{Comfort Factor} = \text{IFD@25\%/IFD@65\%}$$

(7) Report
   a. Report the force in pounds per square inch required for 25 and 65% indentation. These figures are known as the 25% and 65% IFD values, respectively.
   b. Report the length, width and thickness of the specimen.
   c. Report the comfort factor

6.4 INDENTATION FORCE DEFLECTION

(1) Scope
   a. Cellular foam products have traditionally been checked for force deflection by determining the force required to effect a given deflection. Depending on the application, the interest may be in determining how thick the foam is under a given force. In this procedure, the force deflection is determined by measuring the thickness under a fixed force.
   b. This determination shall be known as the indentation force deflection and the measurements as the IFD values.
(2) Apparatus
   a. An apparatus shall be used having a flat circular indentor foot 323 cm^2 in area and equipped with a device for applying forces. It shall be mounted over a level horizontal platform that is perforated with approximately 6.5 mm holes on approximately 20 mm centers to allow for rapid escape of air during the test. The distance between the indentor foot and the platform shall be variable. The apparatus shall be equipped with a device for measuring the distance between plates.
(3) Test conditions
   a. The location of the area for measurement is to be agreed upon with the client. In case a finished part is not feasible for testing, 380 by 380 mm specimens of an average thickness are to be cut from the sample.
   b. The IFD values for molded products are dependent on the specimen dimensions. Difference values are generally obtained for specimens that retain all molded surfaces.
(4) Procedures
   a. Test the whole specimen or a minimum area of 380 by 380 mm. Preflex twice to 60–80% of its full thickness.
   b. Bring the indentor foot into contact and determine the thickness of the specimen.

    c. Indent the specimen until a force of 20 psi is carried by the specimen. Determine the thickness.

    d. Without removing the specimen apply the 40 psi force to the indentor. Determine the thickness.

(5) Report

    a. Report the specimen thickness after 60 ± 3 s at the two (2) psi values. These figures are known as the IFD values, respectively. Also report the length, width and thickness of the specimen.

6.5 COMPRESSION FORCE DEFLECTION

(1) Scope

    a. This test consists of measuring the force necessary to produce a 50% compression over the entire top area of the foam specimen. *Note:* compression deflection tests other than at 50% may be specified as agreed by the purchaser.

(2) Apparatus

    a. An apparatus shall be used having a flat compression foot larger than the specimen to be tested and connected to a force-measuring device. The apparatus shall be arranged to support the specimen on a level horizontal plate that is perforated with approximately 6.5 mm holes on approximately 20 mm centers to allow for rapid escape of air during the test.

(3) Test specimens

    a. The sample shall have parallel top and bottom surfaces and essentially vertical sides. The thickness shall be no greater than 75% of the minimum top dimension.

    b. Specimens from foam shall be a minimum of 2500 mm in area and have a minimum thickness of 20 mm. Specimens less than 20 mm thick shall be plied up, without the use of cement, to a minimum of 20 mm.

    c. The test specimen from molded parts shall have parallel top and bottom surfaces and perpendicular sides. Preferably, the specimen should include both top and bottom molded skins. If a test specimen with parallel top and bottom surfaces including both molded skins cannot be obtained because of the shape of the molded part, at least one of the molded skin surfaces should be retained. Both surface skins should be removed only in cases where the shape of the original sample makes this absolutely necessary.

    d. Maximum molded specimen thickness shall be no greater than the minimum top dimension. Specimens from uncored stock shall have a minimum length of 50 mm, a minimum width of 50 mm, and a minimum thickness of 20 mm. Specimens less than 20 mm thick shall be plied up, without the use of cement, to a minimum of 20 mm.

(4) Number of specimens

    a. Three specimens per sample shall be tested. The value reported shall be the median of those observed. If any value deviates more than 20% from this median, two additional specimens shall be tested and the median for all five values shall be reported.

(5) Procedure

    a. Preflex the specimen twice, 75 to 80% of its original thickness. Then allow the specimen to rest for a period of 6 ± 1 min.

    b. Place the specimen centered on the line of the axial load on the supporting plate of the apparatus.

    c. Bring the compression foot into contact with the specimen and determine the thickness. Compress the specimen 50% of the thickness and observe the final load after 60 ± 3 s.

(6) Report

    a. Report the thickness after contact load and the 50% compression deflection value in pounds per square inch.

6.6 TENSILE TEST

(1) Scope

    a. This test method determines the effect of the application of a tensile force to foam. Measurements are made for tensile strength and ultimate elongation.

(2) Apparatus

    a. The specimen for tension tests shall be a rectangle 10.0 cm long and 1.0 cm wide.

    b. An electronic balance with a bottom hook.

    c. A bracket suitable to hold the foam and suspend it from the hook.

    d. A bracket from which the foam can be pulled (see Figure 36).

(3) Test specimen

    a. The test specimens shall be cut from flat sheet material 10.0 ± 0.1 cm.

(4) Procedures

    a. Test three specimens per sample. The value reported shall be the median of those observed.

    b. Hang the sample from the balance and then slowly pull the sample using the lower bracket, simultaneously watching the readings

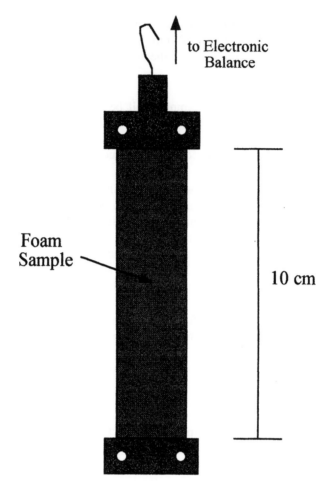

FIGURE 36. Brackets for tensile strength.

on the balance. Pull until the foam breaks. Record the reading on the balance and the length of the foam at break. Measure the cross-sectional area at the break. *Note:* if the force to break the foam exceeds the capacity of the balance, reduce the width of the sample and redo the test.

(5) Calculations

    a. Calculate the tensile strength by dividing the force at break by the cross-sectional area of the specimen.

    b. Calculate the stress by dividing the force at a predetermined elongation by the original cross-sectional area of the specimen.

c. Calculate the ultimate elongation, *A*, by subtracting the original distance between the brackets from the total distance between the brackets at break and expressing the difference as a percentage of the original distance.

$$A\ (\%) = (df - do)/do\} \times 100$$

where:

do = original distance between brackets
df = distance between brackets at the break

d. Report the value as the median value of all specimens tested.
(6) Report the following information
    a. Tensile strength in psi
    b. Stress in psi at a predetermined elongation
    c. Ultimate elongation, in percent

6.7 TEAR RESISTANCE

(1) Scope
    a. This test method covers determination of the tear resistance of foam. The block method, as described, measures the tear resistance under the conditions of this particular test.
(2) Apparatus
    a. The same apparatus used for the tensile strength is used for this test.
(3) Test specimens
    a. The test specimens shall be a block shape free of skins, voids and densification lines. They may be cut on a saw or diecut from sheet material so that the sides are parallel and perpendicular to each other. A 40 mm cut shall be placed in one end as shown in Figure 37.
    b. Three specimens per sample shall be tested. The values reported shall be the median of those tested.
(4) Procedure
    a. Clamp the test specimen in the upper bracket and hang it from the hook on the bottom of the electronic balance.
    b. Pull on the lower bracket while holding the end of the foam perpendicular to the direction of the pull.
    c. Record the force at the point when the foam begins to tear.

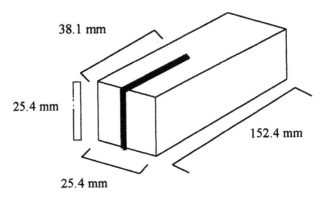

FIGURE 37. Foam sample for tear test.

(5) Calculations

    a. Calculate the tear strength from the maximum force registered on the testing machine and the average thickness of the specimen (direction A-B) as follows:

$$\text{Tear strength, N/m} = F/T$$

where:

F = force in newtons
T = thickness, mm

(6) Report the following information

    a. Tear strength in newtons per millimeter

6.8 AIR FLOW-THROUGH

(1) Scope.

    a. The air-flow test measures the ease with which air passes through a cellular structure. Air-flow values may be used as an indirect measurement of certain cell structure characteristics. The test consists of placing a flexible foam core specimen in a cavity over a chamber and creating a specified constant air-pressure differential. The rate of flow of air required to maintain this pressure differential is the air-flow value

(2) Apparatus
   a. A schematic drawing of the apparatus, including the specimen mounting chamber, pressure gauge, air-flow meters/controller and a vacuum pump, is shown in Figure 38.

(3) Test specimens
   a. The test specimens shall be cut to exceed the size of the screen by a minimum of 1 inch on each side.

(4) Procedure
   a. Place the specimen onto the screen. Make sure that a good air seal is obtained along all edges. The top of the specimen should be flush with the top of the test chamber.
   b. Turn on the vacuum pump.
   c. Adjust the flow rate to achieve a 10 psi pressure.

(5) Report
   a. Report the flow rate at 10 psi or other convenient pressure.

6.9 WATER CAPACITY AND DRAINING

(1) Scope
   a. This test measures the amount of water an undisturbed foam can hold and the amount of moisture held in the matrix of the foam.

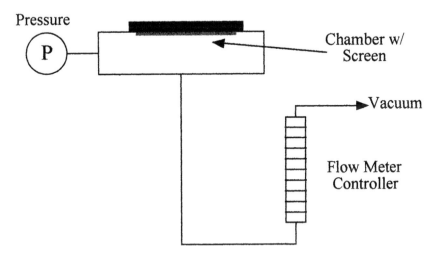

FIGURE 38. Test apparatus for air flow.

(2) Test specimen
   a. A representative specimen of not less than 10 cc in volume shall be cut from a portion free of voids and defects and as near as possible to the section from which the tension and tear specimen will be taken if those tests are to be performed.
(3) Number of specimens
   a. One specimen shall be tested.
(4) Procedure
   a. Determine the mass of specimen in grams to within 10 milligrams [two (2) decimal places].
   b. Immerse the specimen in water at 25°C and squeeze to expel as much air as possible.
   c. Keep the specimen totally immersed in the water for a minimum of 60 minutes.
   d. Using a slotted spoon or similar device, carefully remove the specimen.
   e. Carefully transfer the foam to an electronic balance and record the weight.
   f. Remove the specimen from the balance and hang it from a corner using a convenient device. Allow the foam to drain for 15 minutes. Carefully record its weight.
   g. Squeeze by hand to remove as much water as possible.
   h. Put the specimen in a paper towel and squeeze again.
   i. Weigh the specimen
(5) Calculation
   a. Calculate the water capacity as follows:

$$\text{Water Capacity} = [W_w/W_d] * 100$$

   where:

   W_w = weight of the soaking wet specimen
   W_d = dry weight

   b. Calculate the drained water capacity:

$$\text{Water Capacity} = [W_{wd}/W_d] * 100$$

   where:

   W_{wd} = weight of specimen after draining

   W_d = dry weight

c. Calculate the equilibrium moisture as follows:

$$EM = [(M - D)/M] * 100$$

where:

M = squeezed weight
D = dry weight

(6) Report
   a. Report the water capacity in %.
   b. Report the drained water capacity.
   c. Report the equilibrium moisture in %.

6.10 WICKING

(1) Scope
   a. This test method covers determination of the rate of wicking of foam. The wicking value applies to the immediate area from which the specimen has been taken. It does not necessarily relate to the bulk density of the entire roll or pad.
(2) Test specimen
   a. A representative specimen of not less than 10 cc in volume shall be cut from a portion free of voids and defects and as near as possible to the section from which the tension and tear specimen will be taken if those tests are to be performed. When these dimensions are not possible, a representative portion as agreed upon by the seller and the purchaser shall be used.
(3) Number of specimens
   a. One specimen shall be tested.
(4) Procedure
   a. On a flat surface, place a drop of deionized water on the surface of the foam, being careful not to touch the foam with the dropper, or allowing the drop to fall hard enough to penetrate the surface. In essence, allow the drop to flow from the dropper onto the surface.
   b. Determine the time (in seconds) for the drop to diffuse into the foam.
(5) Report
   a. Report the time in seconds to diffuse.
   b. If the time is immediate, report "less than one (1) second," otherwise report the time.

6.11 COLOR

(1) Scope
    a. This test method covers determination of the degree of yellowing of foam. The color value applies to the immediate area from which the specimen has been taken. It does not necessarily relate to the bulk density of the entire roll or pad.

(2) Test specimen
    a. A representative specimen of not less than 10 cc in volume shall be cut from a portion free of voids and defects and as near as possible to the section from which the tension and tear specimen will be taken if those tests are to be performed. When these dimensions are not possible, a representative portion as agreed upon by the seller and the purchaser shall be used.

(3) Number of specimens
    a. One specimen shall be tested.

(4) Procedure
    a. Under standard fluorescent lighting, compare the color of the foam with the color chart developed as follows: a column is made with a vivid-orangish yellow (hue = 52°, saturation = 100 and lightness = 75). A gradient is applied from white to vivid orangish yellow (lightness = 80 to 100). The column is divided into 10 equal segments. This first segment on the white end is 1 and the darkest is 10.

(5) Report
    a. Report the color.

6.12 LINEAR SWELL

(1) Scope
    a. This test method covers determination of the linear swell of foam. The swell value applies to the immediate area from which the specimen has been taken. It does not necessarily relate to the bulk density of the entire roll or pad.

(2) Test specimen
    a. A representative bone-dry specimen of not less than 10 cc in length shall be cut from a portion free of voids and defects and as near as possible to the section from which the tension and tear specimen will be taken if those tests are to be performed. When these dimensions are not possible, a representative portion as agreed upon by the seller and the purchaser shall be used.

(3) Number of specimens
   a. One specimen shall be tested.
(4) Procedure
   a. Measure the length of a specimen of foam to the nearest 0.1 cm.
   b. Immerse the sample in water at 25°C.
   c. Allow the foam to soak for a minimum of 10 minutes.
   d. Remove the sample from the water and wring dry in a paper towel.
   e. Measure the length of the specimen.
(5) Calculation

$$\% \text{ Linear Swell} = ((\text{wet length} - \text{dry Length})/\text{dry length}) * 100$$

(6) Report
   a. Report the linear swell.

CHAPTER 7

Quality Systems Regulations

7.1 INTRODUCTION

We have been discussing a product development structure within which a device can be developed. The structure begins with the design of a product. This definition is translated into a formulation and a process by which the product can be.made into a functioning commercial product. This development process includes a confirmation that the product is safe and effective as validated by a series of analytical procedures many of which we covered in the previous chapter.

It could be argued that the process is completed with these three steps. Indeed, until recently, that was the process by which products were brought to market. However, another step has been added. In the case of medical devices, the design and manufacture of the product must be conducted inside a protocol that has come to be known as a quality system. It was formerly known as good manufacturing practice (GMP) for production and good laboratory practice (GLP) for testing. Recently, however, the U.S. Food and Drug Administration Good Practice Regulations, CFR Part 820, included the monitoring of the design or development process. This has been combined into a series of regulations known as quality system regulations (QSR).

Internationally, the philosophy of establishing a system under which organizations should operate has focused on ANSI/ASQC Q9001-1994 guidelines known as ISO 9000 [21]. The purpose of this blueprint is to promote consistent quality. Note that the system doesn't seek to improve the quality per se, but rather to make the quality consistent and therefore predictable.

It is not the purpose of this book to describe the systems and guidelines

119

under which a development and production effort should be conducted. It has become increasingly important that an organization undertaking the development of a new product seek the assistance of a professional in regulatory affairs to assist with the process. This is especially true, one might even say required, of a medical device developer. Outside the medical arena, if a company has or seeks registration as an ISO 9000 company, a QSR is required. We will return to the product development paradigm we have been using, but modify it to reflect this new requirement.

The significance of redefining the paradigm as shown in Figure 39 is to emphasize the integral nature of the co-development of a product and the quality system within which it will be manufactured.

As researchers in medical devices, we are in a state of flux vis-à-vis the system under which we must operate. On the one hand, in the United States we are under the direction of the FDA. On the other hand, as global entrepreneurs we must also do our work in conformance with the international standards (ISO 9000). Fortunately, there is a strong driving force towards reconciliation between the ISO and QSR. Emphasis on the individual items in the guidelines is different in the systems. The design requirement of the U.S. QSR is more organized than in ISO. In general, however, the systems are close enough, at least for this discussion, and we can consider them equivalent in the following sense. If a company rigorously follows the proper ISO 9000 guideline, it will also be fulfilling the FDA's QSR requirements.

For this discussion, therefore, we will use the ISO 9000 guidelines. ISO 9000 is a series of guidelines that reflect the practice and purpose of manufacturers. We have conducted our discussion in the context of a company

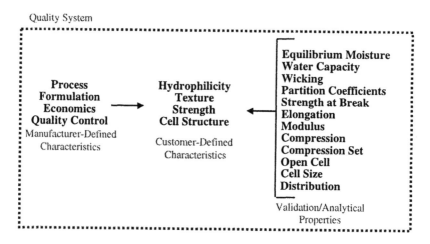

FIGURE 39. Product development within a quality system.

or an individual that seeks to design a new product. In this case, ISO 9001 is the proper standard. ISO 9002, by comparison, deals with companies that receive instructions and/or drawings from a client and manufacture a product without any modifications. This is typical of the relationship a client has with a machine shop. Other standards deal with companies that are basically marketers of other companies' products. It is interesting to note, and in a sense the purpose of this book, that a company maintains design control of a device. With the information herein, the company should have enough information to guide the research of the foam producer with sufficient skill so that the producer's contribution is minimal.

The full name of a set of standards known as ISO 9000 is ANSI/ASQC Q9000-1-1994 [21]. It is composed of four sections. They are:

- scope
- reference
- definitions
- quality system regulations

The scope defines what sections of a facility will be covered under the quality system. The reference section defines the regulations and guidelines under which the plant will be controlled. The definition section seeks to standardize the terms that are used in the system. Typically, this section will reference another ANSI/ASQC document where accepted definitions are listed.

The fourth section is the core of the document. It is separated into 20 subsections. The differences between the various standards mentioned above (9001 vs. 9002) are which of the 20 subsections are covered. For instance, 9001 includes all the subsections while ISO 9002 does not include subsection 4.4, the guidelines that control the design of new technologies.

Again it is not the purpose of this book to teach QSRs. There are points to be made, however, with regard to specific subsections of the document. In those subsections, the design and manufacture of products based on hydrophilic polyurethane have certain implications, which we will discuss.

Before we get to that, however, another subject deserves attention. It concerns the regulations in the United States that define the roles of the participants in the development cycle. For an explanation, an individual can contract with a manufacturer to make a device. The role of that individual can range from being a "specification writer" to the actual manufacturer (even though he or she doesn't physically make the device). The status with the FDA is defined in a document submitted to the FDA, which establishes the individual's role in the production cycle. As a "spec writer," the company is only responsible for supplying someone else in the chain with a set of requirements. Other members of the chain are responsible for the development and/or the manufacture. An individual might, alternatively, decide to subcontract the manufacture of the device but register with the FDA as the manufacturer. In still another variation, an individual or company might define a device and relinquish or maintain control of the design process. If the design process is kept internally, the company would be required to register as an ISO 9001 company. By relinquishing control, an ISO 9002 registration might be in order. In either case, the FDA has a right to examine the design process, either directly or indirectly.

In summary, and taking a holistic view of the process, all of the segments of the ISO regulations and the QSR must be addressed by someone in the design/manufacture structure. Who does what is a matter of convenience, resources and talent. It is therefore useful to describe the 20 subsections and the various perturbations described above. Following that we will go into detail about the subsections that have specific implications for the design and manufacture of hydrophilic polyurethane products.

Rather than taking a classic view of the subsections, we have chosen to develop what we feel is a functional view of the regulations. All of the subsections are covered by this view, but they are arranged to reflect the differing roles a developer of products can assume (spec writer, subcontractor, etc.) (see Figure 40).

This is not the place to describe each of these blocks: suffice it to say they cover all of the practices of a company with the exception of finance. You will notice that only 18 of the 20 subsections of the standard are included in the above chart. Missing is the subsection on service, which is not typically a function of manufacturers in this business. Also not shown in Figure 40 is the requirement for the organization to submit itself to internal audits. The purpose of Figure 40 is to illustrate the activities that an oragnization must engage in to properly develop and manufacture a device. It is assumed that an internal audit will be practiced.

The essential purpose of the system is that an environment for consistent quality is created by management and defined in the quality manual (QM). The QM establishes certain requirements for the development of contracts with clients and suppliers and specifies that the manufacture of products be documented in terms of record keeping, training and documentation of the results.

The roles of the individuals in design and manufacture of a new product, as we described above, are shown in the next few figures. Remember that all of these guidelines must be the responsibility of someone in the design/production system.

In our experience the most common arrangement involves a product designer and marketer without manufacturing capabilities. In this case, the

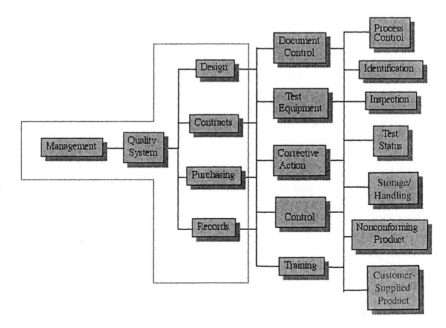

FIGURE 40. Operational view of ISO 9001 regulations.

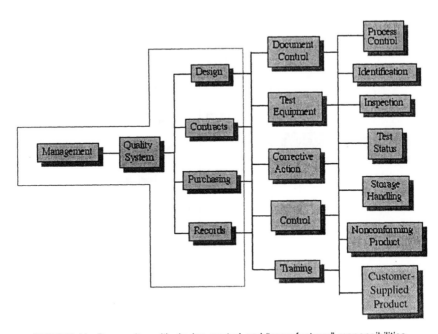

FIGURE 41. Spec writer with design control and "manufacturer" responsibilities.

123

designer maintains design control and lists with the FDA (if it is a medical device) as the manufacturer. This is represented by Figure 41.

In this diagram, the overall management responsibilities are maintained, as well as contract development and control of the production records. The relationship with the manufacturer is documented in the contract that the contractor will develop. Included, as we will discuss, are assurances that the subcontractor will adhere to certain levels of quality. These, at a minimum, include safety and efficacy data, but extend to training, etc. Thus, during an audit (FDA or otherwise), the spec writer/designer/manufacturer will be able to demonstrate, with objective evidence, that the product is made under conditions of uniform quality.

As a contract manufacturer, the responsibilities shift to the right side (see Figure 42). Note, if the manufacturer chooses not to register as an ISO 9000 class facility, the above requirements are set by the owner of the project as the minimum needed in order to be the manufacturer of the product. In cases where the manufacturer makes only a portion of the final device (a component), the company, at the present time, does not even

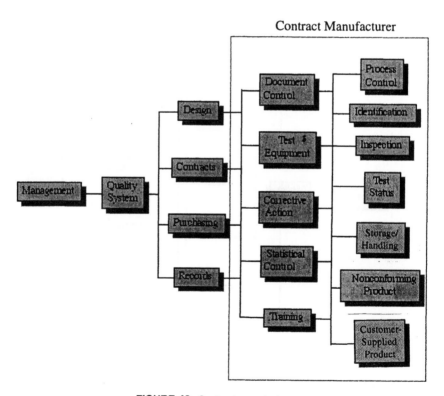

FIGURE 42. Contract manufacturer.

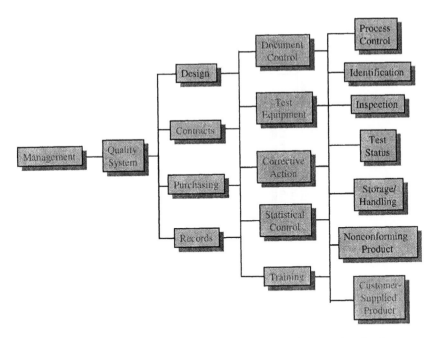

FIGURE 43. Subsections of concern to hydrophilic polyurethane product developers.

have to operate under a legal requirement for a QSR. The contractor, however, is required to develop a contract that ensures that the product is made as if it were under ISO 9000 control.

In practice, the bulk of the work in maintaining QSR or ISO 9000 status falls upon the manufacturer. The design subsection includes items and practices that are affected by the peculiarities of hydrophilic polyurethane.

The remainder of this chapter focuses on the subsections that deal with the design and physical manufacture of hydrophilic polyurethane products. Only those subsections that are affected by the fact that we are using hydrophilic polyurethane will be covered. These are highlighted in Figure 43.

Each of these subjects will be covered starting with the text of the first lines of the ANSI/ASQC Q9001–1994 guidelines [21] and suggestions as to how to approach the subject from this perspective.

7.2 DESIGN CONTROL

The supplier shall establish and maintain documented procedures to control and verify the design of the product in order to ensure that the specified requirements are met [21, p. 3].

126

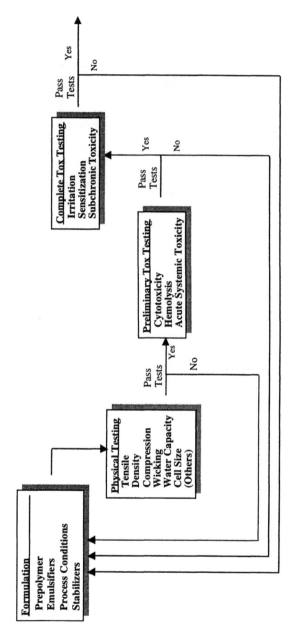

FIGURE 44. Four step program for the design and testing of a medical device.

Throughout this book we have worked with a product development paradigm that fulfills many of the requirements of this section. From a practical point of view, the process starts with a definition of the product. This is followed by a translation of the requirements into the formulations and processes used with hydrophilic polyurethanes. It goes without saying that this phase is an interactive process between the designer and the production unit. The importance of this section is that there is a documentation component to the process.

Throughout, the QSR systems require that the parties involved be able to submit objective evidence that the procedures were followed. This is especially true of the design subsection in as much as government units can get involved.

In this instance, the objective evidence can be meeting notes where the design requirements are defined and revised. This does more than provide the objective evidence to the regulatory body. In our experience it minimizes the chances for misunderstandings.

The need for objective evidence in the form of written materials is most important in the validation of the product (Does it meet the design requirements?), and in the validation of the process (Can it consistently make in-spec product?).

We developed a graphic (Figure 44) that describes a system by which a new medical device can be developed. It is related to the product development paradigm we have been using, but concentrates more on the dual responsibilities of efficacy and safety (toxicology). In this case, the scheme designed includes a specific battery of toxicity tests that must be passed. By adding or subtracting tests, the program can be modified to fulfill the requirements of other products.

Following the initial conversion of the design into samples, the least expensive toxicity testing is conducted. If any of the tests is failed, the development process returns to the formulation development stage and physical testing. It is not until the initial tests have been passed (both efficacy and safety) that the investigation advances to the more expensive testing.

7.3 PROCESS CONTROL

The supplier shall identify and plan the production, installation and servicing processes which directly affect quality and shall ensure that these processes are carried out under controlled conditions [21, p. 5].

During the design phase of the development process, the important product characteristics will have been identified. Along with the size, the porosity, cell size and the compression of the device were described. The process development phase of the project determines the process parameters that affect these characteristics. Further, it is the validation of the process that determines the ranges of the process parameter that can yield an acceptable product. Thus upon completion of the project, all of the parties in the venture should have knowledge of the critical process parameters. These must then be translated into a document that represents the work instructions. The documentation will also include the necessary archiving of data to allow the issuing of a certificate of analysis. This testifies that the analysis of the material is within the design specifications. When confidentiality of the process is critical, a certificate of compliance might be in order. This document is useful from a record-keeping point of view in that it testifies that the producer has manufactured the product under the agreed-upon process conditions.

By way of example, a product might require careful control of the density. During the process development stage, it was determined that the most critical factor in the control of density was the temperature of the aqueous/prepolymer emulsion. While the certificate of analysis will list the density, the procedures established to fulfill the requirements of an ISO 9000 registration would require that the emulsion temperature be measured.

Another aspect of the process control requirement of the standard is the organizational flow. This is not specifically required, but in practice it is a good idea to develop a process flow-chart that describes the sequence of events during the production of a device. For example, the process to manufacture roll stock of a medical grade of hydrophilic polyurethane foam is described by the flowchart in Figure 45.

There are several advantages to this approach in process control.

- *Audits:* it is generally true that a clear and concise description of a process is necessary. This is greatly aided by a flow-chart. This is especially true if the in-process analysis and the disposition of non-conforming material is included
- *Training:* the chart can be used as a control mechanism to ensure that each step of the process is covered by an employee who is trained. Further, part of the requirements of the ISO 9000 guidelines is that employees clearly understand their function and responsibilities. We have found it advantageous to display the flowchart and to periodically test people to ensure they are clear and concise concerning their function.
- *An index for the work instruction manual:* in each of the boxes, the number of the work instruction associated with a given activity can

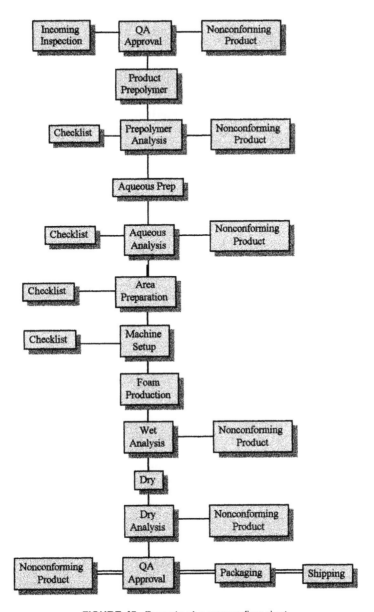

FIGURE 45. Example of a process flow chart.

be placed. By this technique, employees not only know where they stand in the process, they also have a reference number of the written instructions associated with their current activity.

- *Identifying critical steps in the process:* by any one of several techniques, the critical portions of the process can be highlighted. This is particularly useful in describing the process to auditors by demonstrating that you have reviewed and are aware of where special attention needs to be concentrated.
- *Assigning trained individuals:* in planning the production of a new product that doesn't require new procedures, a chart of this type is useful in assigning trained individuals with assurance and, more importantly, documentation that the manufacturing will be conducted by qualified individuals.
- *Labor costing:* by imbedding labor and capital cost in the chart, a convenient way to cost a process is developed.

In summary, this section of the standard requires that the manufacturer be able to produce objective evidence that the material was produced in accordance with the agreed-upon level of process control.

7.4 IDENTIFICATION AND TRACEABILITY

. . . the supplier shall establish and maintain documented procedures for identifying the product by suitable means from receipt and during all stages of production, delivery and installation [21, p. 5].

The manufacture of products based on hydrophilic polyurethane technology involves both continuous and batch processes. It is required by the standard that the manufacturer be able to identify the status of the material at all stages of manufacture. The producer must assign identifying numbers to maintain the integrity of the documentation system. Especially for medical devices, the manufacturer is required to trace the components of a product back to the raw materials. In the case of a hydrophilic polyurethane manufacturer, this means that the lot number of the prepolymer be known and that the identification or tracking system include references. If the manufacturer is also the maker of the prepolymer, the tracking system might go back to the raw materials that went into the prepolymer.

No standard gives guidance as to how these identification numbers are developed. A typical way is to use the date of manufacture. This has the advantage that it can represent a full day of production with only one start-up.

Alternatively, the lot number of the polymer can be used to identify a production run. The batches can be as much as 3000 lbs., and are arguably the most important independent variable. Conversely, if the aqueous batches are large, the aqueous could serve as the basis for identification.

In any case, it is required that the manufacturer develop, implement and maintain a system that identifies the product at all stages of production. This system includes assurances that material that does not conform to the specifications are clearly identified as such. This is to ensure that the material does not find its way to the shipping department and is not accidentally delivered to the customer.

7.5 INSPECTION AND TESTING

The supplier shall establish and maintain documented procedures for inspection and testing activities in order that the specified requirements for the product are met [21, p. 6].

In Chapter 6, we covered many of the methods by which the product is to be tested. While these tests are oriented toward laboratory results, in practice, however, it is often possible to show conformance with the design specifications by using less sophisticated methods.

An important part of this section is the recommendation that materials be tested at various stages during the manufacturing process. This has an economic component. If a material is found to be nonconforming at an early stage of production, the manufacturer has the option of removing it from the production flow so subsequent value-added steps are not performed to a product that will ultimately be rejected.

In-process analysis also permits trend analysis for process control reasons. It is also a useful tool in defining production problems and opportunities for evolutionary process development.

7.6 STORAGE AND HANDLING

The supplier shall establish and maintain documented procedures for handling, packaging, preservation and the delivery of product [21, p. 8].

For the most part, conventional care is sufficient to at least prevent physical damage. Two factors need to be taken into account in establishing the procedures to fulfill the requirements of this standard.

First, the chemistry of hydrophilic polyurethanes, due to an undefined characteristic, has the ability to adsorb certain hydrocarbons on its surface. Therefore, if the material is stored in an environment that has high levels of hydrocarbons, like from a gasoline or LPG forklift truck, the hydrophilic polyurethane will become contaminated. A yellowing of the foam evidences this.

Secondly, the most common hydrophilic polyurethanes are produced using toluene diisocyanate or other aromatic isocyanate. Aromatic compounds in general are subject to degradation by the action of ultraviolet light. The reaction causes a ring opening that results in yellow-colored bodies. Again, the evidence is the yellowing of the foam. This difficulty can be mitigated by the use of ultraviolet stabilizers, but the problem cannot be eliminated.

Because of these factors, it is necessary that the foam be stored:

- away from possible hydrocarbon contamination
- protected from light

7.7 NONCONFORMING PRODUCT

> The supplier shall establish and maintain documented procedures to ensure that product that does not conform to the specified requirements is prevented from unintended use or installation [21, p. 7].

It is said that a roll of hydrophilic polyurethane foam, regardless of its size, is a single molecule. This comment is meant to reflect that there is sufficient crosslinking to assume there are no monomers or oligomers in the foam. This is not unreasonable for practical purposes and is supported by the low extractables in a well-fabricated foam. The molecular structure, therefore, is that of a gel, a hydrogel, if you will. As such it is not thermoplastic in the conventional sense. (It can be heat-sealed, however, but this is because at high temperature, the polymer degrades to a liquid, which upon cooling, solidifies. This gives it the false appearance of being thermoplastic.)

For the most part, this gel effect is to our advantage. Indeed, the foaming process depends on the fact that the molecule crosslinks. Without that effect we would end up with elastomers instead of foams.

The difficulty with this, however, is that reworking is typically not possible. Thus, the alternative to a nonconforming product is disposal or respecification. To illustrate we will return to the model process we used in Chapter 4 in this book (see Figure 46).

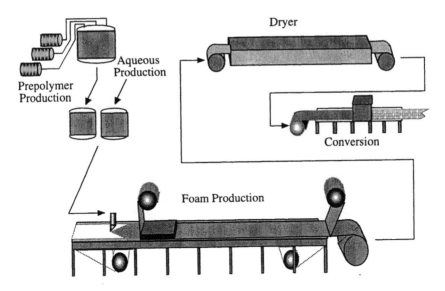

FIGURE 46. Continuous process to make hydrophilic polyurethane foam.

7.7.1 Prepolymer Production

If the manufacturing scheme includes the production of the prepolymer, analytical requirements will stipulate that the NCO concentration be determined and perhaps the viscosity. Other analyses are possible and may even be advisable, but this is probably the minimum. In none of the processes that are currently used is it possible to correct or reprocess a prepolymer even if the NCO level is below specification, in as much as the NCO includes both free TDI (for instance) and end group NCOs.

Therefore, back-adding isocyanates is not an acceptable remedy. Thus if a prepolymer is out of spec, it must be disposed of. With appropriate validation, a procedure to blend off nonconforming product can be devised.

7.7.2 Aqueous Production

This component of the formulation is typically inexpensive both in raw materials and in labor costs. Disposal is usually advised.

7.7.3 Foam Production

As we discussed, reprocessing the foam to bring it into conformance with specifications such as size and density, cell size, and absorption char-

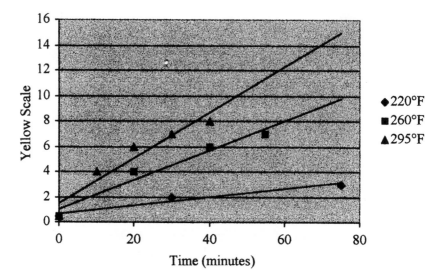

FIGURE 47. Yellowing as a function of temperature.

acteristic is all but impossible. Thus, once the foam is made and found not in conformance with the specifications, remedies are limited to two categories. The foam might be respecified into another use where the product would be in spec. Apart from that, the alternative is disposal.

7.7.4 Drying

This unit operation presents one of the few possibilities for remediation. If the foam does not come out dry, it can usually be redried. Care must be taken to avoid a situation in which the edges of a product dry before the body. Upon re-drying, the edges might experience temperatures that would cause them to yellow. The graph in Figure 47 shows the effect of temperature on yellowing. It uses the yellowing chart in Chapter 6, Section 6.11.

7.8 CONVERSION

With careful inspection, most nonconformities will be identified before this step in the process. Examples from this step are typically miss-struck pieces or incomplete cutting. In either case the product is typically discarded.

REFERENCES

1. Woods, G. *The ICI Polyurethanes Book*, John Wiley & Sons, 1987.
2. Russel, D. *Practical Chemistry of Polyurethanes and Diisocyanates*, Akron Polymer Laboratory, 1992.
3. Woods, G. ibid.
4. ibid.
5. Russel, D. ibid.
6. Bergstrom, N. et al. *Pressure Ulcer Treatment*, No. 15, Rockville, MD: U.S. Department of Health and Human Services, Public Health Service, Agency for Health Care Policy and Research, AHCPR Pub. No. 95–0653, Dec. 1994.
7. ibid.
8. Winter, G. D. Formation of Scab and Rate of Epithelialization of Superficial Wounds in the Skin of Young Domestic Pigs, *Nature*, 193:293–294, 1962.
9. U.S. Wound Management Markets. *The Quest for Less Damaging, More Active Products*, Frost & Sullivan, Inc., 1996.
10. Pflugmacher, U. and Gottschalk, G. *Appl. Microbiol. Biotechnol.*, 41:313–316, 1994.
11. Bailliez, C., Largeau, C., Casadevall, E., Yang, L. W. and Berkaloff, C. *Appl. Microbiol. Biotechnol.*, 29:141–147, 1988.
12. Hu, Z-C., Korus, R. A. and Stormo, K. E., *Appl. Microbiol. Biotechnol.*, 39:289–295, 1993.
13. Braatz, J. A., Heifetz, A. H. and Kehr, C. L. *J. Biomater. Sci. EDN.*, 3(6):451–462, 1992.
14. Merrill E. W. and Salzman, E. W. *J. Am. Soc. Art. Int. Org.*, 6:60, 1983.
15. Nagaoka, S. and Mori, Y., et al. *J. Am. Soc. Art. Int. Org.*, 10:76, 1987.
16. Jeon, S. I., Lee, J. H., Andrade, J. D. and DeGennes, P. G. *J. Colloid Int. Sci.*, 142:149, 1991.
17. Jeon, S. J. and Andrade, J. D. *J. Colloid Int. Sci.*, 142:159, 1991.
18. Wood, L. L. US Patent 4,158,087.
19. Eagan, J. Personal Communication.
20. Available from the American Society for Quality Control, 611 East Wisconsin Avenue, Milwaukee, WI 53202.
21. Marans, N. S. et al. US Patent 4,132,839.

T - #0229 - 111024 - C0 - 234/156/7 - PB - 9780367398637 - Gloss Lamination